THE SCIENTIFIC REVOLUTION AND OF MODERN SCIENCE

Studies in European History

General Editor: Richard Overy
Editorial Consultants: John Breuilly
                        Roy Porter

PUBLISHED TITLES

FORTHCOMING

The Scientific Revolution and the Origins of Modern Science

John Henry
Senior Lecturer in the History of Science
University of Edinburgh

First published in Great Britain 1997 by
MACMILLAN PRESS LTD
Houndmills, Basingstoke, Hampshire RG21 6XS and London
Companies and representatives throughout the world

A catalogue record for this book is available from the British Library.

ISBN 0–333–56047–7

First published in the United States of America 1997 by
ST. MARTIN'S PRESS, INC.,
Scholarly and Reference Division,
175 Fifth Avenue, New York, N.Y. 10010

ISBN 0–312–16540–4

Library of Congress Cataloging-in-Publication Data
Henry, John.
The scientific revolution and the origins of modern science / John
Henry.
p.   cm. — (Studies in European history)
Includes bibliographical references and index.
ISBN 0–312–16540–4 (pbk.)
1. Science—History.   2. Science—Methodology.   3. Science–
–Philosophy.   I. Title.   II. Series: Studies in European history
(New York, N.Y.)
Q125.H5587   1997
509—dc20   96–27605
                                                    CIP

© John Henry 1997

This book is printed on paper suitable for recycling and made from fully managed and sustained forest sources.

10   9   8   7   6   5   4   3   2   1
06   05   04   03   02   01   00   99   98   97

Printed in Hong Kong

For my sister, Kay (1948–1996)
In Memoriam

Contents

A Note on References

References are cited throughout in brackets according to the numbering in the select bibliography, with semi-colons separating each item. Page references, where necessary, are indicated by a colon after the bibliography number.

Editor's Preface

The main purpose of this series of studies is to make available to teacher and student alike developments in a field of history that has become increasingly specialized with the sheer volume of new research and literature now produced. These studies are designed to present the 'state of the debate' on important themes and episodes in European history since the sixteenth century, presented in a clear and critical way by someone who is closely concerned with the debate in question.

The studies are not intended to be read as extended bibliographical essays, though each will contain a detailed guide to further reading which will lead students and the general reader quickly to key publications. Each book carries its own interpretation and conclusions, while locating the discussion firmly in the centre of the current issues as historians see them. It is intended that the series will introduce students to historical approaches which are in some cases very new and which, in the normal course of things, would take many years to filter down into the textbooks and school histories. I hope it will demonstrate some of the excitement historians, like scientists, feel as they work away in the vanguard of their subject.

The format of the series conforms closely with that of the companion volumes of studies in economic and social history which has already established a major reputation since its inception in 1968. Both series have an important contribution to make in publicising what it is that historians are doing and in making history more open and accessible. It is vital for history to communicate if it is to survive.

R.J. OVERY

Acknowledgements

As a number of my colleagues know, this little book took me far longer to produce than it should have done. It gives me great pleasure, at last, to be able to thank Richard Overy and Roy Porter for their unfailing patience and encouragement. At Macmillan, first of all Vanessa Graham, and then Simon Winder, also proved endlessly encouraging and unbelievably forbearing as one deadline after another went by. I am very grateful to both of them for this generous treatment. The author's greatest debt when working on a book of this kind, however, is to the authors of all those books and articles upon which he drew in the writing of it. The full list of all these would be longer than the Bibliography upon which this book is ostensibly based. Thanks to these authors I can wholeheartedly concur with Robert Hooke that 'the more you are informed, the more able you are to inquire for and seek after what is considerable to be farther known concerning that Subject'. I very much hope that readers of this book will also feel themselves more able to seek further for a fuller understanding of the origins of modern science. Throughout the writing of this book my sister, Kay, was bravely enduring the ravages of breast cancer and the medical treatment intended to assuage it. She died a matter of weeks before I was able to complete it. We always loved each other very dearly and I dedicate this book to her memory.

JOHN HENRY
Edinburgh, 1996

1 The Scientific Revolution and the Historiography of Science

The Scientific Revolution is the name given by historians of science to the period in European history when, arguably, the conceptual, methodological and institutional foundations of modern science were first established. The precise period in question varies from historian to historian, but the main focus is usually held to be the seventeenth century, with varying periods of scene-setting in the sixteenth and consolidation in the eighteenth. Similarly, the precise nature of the Revolution, its origins, causes, battlegrounds, and results vary markedly from author to author. Such flexibility of interpretation clearly indicates that the Scientific Revolution is primarily a historian's conceptual category. But the fact that the notion of the Scientific Revolution is a term of convenience for historians does not mean that it is merely a figment of their imaginations with no basis in historical reality.

Certainly, knowledge of the natural world can easily be seen to have been very different in 1700 from the way it was in 1500. Undoubtedly, during this period, highly significant and far-reaching changes were brought about in all aspects of European culture concerned with the nature of the physical world and how it should be studied, analysed and represented, and many of these developments continue to play a significant part in modern science. The concept of the Scientific Revolution can be seen, therefore, to refer to a very real process of fundamental change. If we wish to understand the nature and causes of these changes we must try to pin-point the fundamental issues for past thinkers, their most significant switches in ways of thinking, the clearest shifts in their social organization, the most far-reaching changes in their scientific practice, and the implications of the most significant discoveries and inventions. We need not detain ourselves, however, in discussions about the correct starting date, about precisely what kind of revolution it was, or about the best way of defining revolutionary change in science. To do so is mistakenly to regard what is nothing more than a convenient term of reference for a wide range of major social and intellectual changes as though it somehow grasps the putative essence of those changes [32; 170].

The reification of the Revolution, as a revolution, has, however, given rise to one important historiographical debate; one which continues to be disputed. A number of historians have argued that the very concept of a revolution in early modern science, with its implication of a radical break with the past, is misplaced or misconceived. The issue depends, of course, entirely upon whatever criteria are used to circumscribe the debate. The current consensus seems to be that the 'continuist' view of scientific development has been overstated in the past, but remains valuable for pointing to the many and various antecedents of later developments during the medieval period [139]. Where once the Middle Ages could be presented as a period of scientific sterility and stagnation, thanks to the excellent work of continuist historians, we can now see the undeniable achievements of medieval thinkers, particularly in the fields of astronomy and cosmology, optics, kinematics, and other mathematical sciences, as well as in the development of the notion of natural laws and of the experimental method [41; 42; 173].

Moreover, continuist historiography has played an important part in making historians of science aware of the dangers of what is called whiggism. There is a tendency in the history of science to look back with hindsight about what is known to be important later. To judge the past in terms of the present is to be whiggish. In the early decades of the formation of the discipline it was common for a historian of science to pick out from, say, Galileo's work, or Kepler's, those features which were, or could most easily be made to look like, direct anticipations of currently held science. The resulting history was often a lamentable distortion of the way things were. But the very notion of the Scientific Revolution, it's easy to see, has something rather whiggish built into it. The science of that era was revolutionary because, unlike previous science, it was like our own, or so we think. It's almost as if what we want to say is not just, here are the origins of modern science, but here is the beginning of current science.

There is a sense in which this kind of whiggism still thrives in history of science. The *raison d'être* of history of science is, essentially, to try to understand why and how science became such a dominant presence in our culture [161]. As such, all our history is directed towards the present. So, although the vigorous repudiation of whiggism has now become a shibboleth which must be uttered to gain entry into the ranks of serious scholars, whiggism lurks within all of us [98]. The distinguished intellectual historian Richard H. Popkin once wittily announced that he intended to study the reasons why 'one of the greatest anti-Trinitarian theologians of the seventeenth century', Isaac Newton (1642–1727), should take time off to write works on natural science [169]. Why do I describe this as witty? Because I find it impossible to take seriously the suggestion that Newton's historical significance derives from his standing as a theologian. To this extent I have to confess to whiggism. I believe that we study Newton because he made such exceptional contributions to our scientific culture. Nothing else about this fascinating man is quite so interesting as that.

Continuism can be seen, however, at least to some extent, as an antidote to whiggish tendencies because it tends to be backward-looking, rather than inherently forward-looking. The attempt to see Galileo (1564–1642) as a latter-day *impetus* theorist is intrinsically less whiggish than presenting him as one who prefigured Newtonian inertia, and should lead us closer to Herbert Butterfield's suggestion (before he himself became a Whig historian by writing a book about the Scientific Revolution! [98: 58]) that we should attempt 'to see life with the eyes of another century than our own' [quoted in 98: 48]. Virtually all historians of science now try to avoid overt forms of whiggism, revolutionists and continuists alike, but it seems safe to say that historians of the Middle Ages were among the first to show the way [41; 42; 55].

Another indicator that the concept of the Scientific Revolution is inherently whiggish is the very word 'scientific'. Our present use of the word 'science' was first coined in the nineteenth century and, strictly speaking, there was no such thing as 'science' in our sense in the early modern period. To talk as though there was, as I have been doing, is an obviously whiggish distortion. Part of our aim, in looking at the historical development of what we think of as science, should be to understand how the very concept 'science' arose; we simply beg the question if we talk about 'science' as though it always existed.

So, if there was no 'science' at the time of the Scientific Revolution what was there? There was something called 'natural philosophy', which aimed to describe and explain the entire system of the world. There were a series of technically developed disciplinary traditions, either mathematically based like astronomy, optics, mechanics and kinematics and what was called music, but which we would see as a rather more mathematical study based on principles of ratios and other aspects of proportion [30; 34]; or medically based like anatomy, physiology, and pharmacology [76; 77; 101; 165; 226]. And finally, there were a range of practical arts like navigation, cartography, fortification, mining, metallurgy, and surgery [193: 225–7; 136]. The relationship between these technical disciplines and natural philosophy requires careful elucidation and work in this area is continuing [80; 81].

Some of the most exciting research in the history of science has been concerned to show how changing interactions between the specialist disciplines and natural philosophy, through practitioners in either or both camps, has given rise, not only to new developments in knowledge and practice, but also to something which looks closer, or more directly related, to our present day demarcation of scientific disciplines. Galileo's endeavour to bring together kinematics and natural philosophy resulted in what he called the new science of motion, which historians still regard as an influential step towards subsequent theories [205]. Similarly, the new and highly influential natural philosophy of René Descartes (1596–1650), the mechanical philosophy, was forged out of his attempts to base natural philosophy upon the certainties of geometrical reasoning [88]; and Newton's new natural philosophy was based, as the title of his book proclaimed, on mathematical principles [80]. The development

4

of atomistic theories of matter grew at least partly out of the efforts of medically trained natural philosophers to extend Aristotle's natural philosophy to account for the empirical knowledge of chemists [63]. The new experimental philosophy, developed in late seventeenth-century England by Robert Boyle (1627–91) and others, was intended to demarcate new discipline boundaries around correct natural philosophy, and to exclude what had previously been regarded as correct practice [201].

It should be clear from this that it is by no means ideal to use the term 'natural philosophy' instead of 'science' when dealing with the early modern period. The terms are by no means equivalent. One of the revolutionary things about the Scientific Revolution is precisely the fact that, throughout the period, natural philosophy was being changed beyond all recognition, and coming closer to our concept of science. Even so, the term natural philosophy was the one which was most used at the time, to refer to an understanding of the physical world. Accordingly, I will use 'natural philosophy' and 'science' quite interchangeably, meaning in both cases nothing more than the endeavour to understand, describe or explain the workings of the physical world (I will also use the adjectival forms, 'natural philosophical' and 'scientific' in the corresponding way). I hope neither anachronism will prove too distracting.

It is possible to acknowledge a whiggishness in one's reasons for looking at the history of science without allowing whiggism to intrude into our historical narratives. Rather than imposing our own views, our aim as historians should be to strive for as full an understanding as possible of the contemporary context. For example, if we wish to understand the contemporary response to a little book like Galileo's *Siderius Nuncius* (*Starry Messenger*, 1610), in which Galileo presented the discoveries he had made by turning the newly invented telescope to the night sky, it is obvious that we cannot simply read Galileo's text [82]. Nor will it be enough to familiarize ourselves with the technical astronomy and cosmology of Galileo's time. It is well known, for example, that some of Galileo's contemporaries refused to look through his telescope. Why did they respond that way? Obviously not because of any astronomical technicality [220]. An understanding of the implications of what Galileo wrote is

5

also relevant. A contemporary reader would have responded differently to a work which was likely to have no impact beyond its immediate area, than to one which could be seen to run counter, not only to current astronomy and cosmology, but also to the wider natural philosophy, and to religious belief. A really full account would also require some knowledge of contemporary understanding of who Galileo was: his reputation, his presumed or perceived motivation, and whether he could be held to speak disinterestedly, and so forth [13; cf., for similar considerations about other thinkers, 201; 200; 198]. Of course, there are no definite limits to an enterprise of this kind, which is why no single historian can have the last word on any given topic. It is always possible to point to something else in the subject's milieu which may be relevant to our attempt to reconstruct the past.

A striving for an ever richer contextualization can be seen, then, as the driving force in current historiography of science. This contextualism can be seen as the outcome of an eclecticism which combined what used to be seen as opposite approaches to the history of science. The discipline of the history of science used to be riven by warfare between internalists and externalists (c. 1930–59). The internalists were supposed to have believed that science, or possibly an individual subdiscipline within science, was a system of thought which was self-contained, self-regulating, and developed in accordance with its own internal logic. The externalist, on the other hand, was supposed to believe that the development of science was determined by the socio-political or socio-economic context from which it emerged. In fact, neither position seems to have been properly established as valid or viable [199: 345 51], and it wasn't long before a professed eclectic approach became all the rage (c.1960). Effectively, this eclectic approach is still dominant, and what this means in practice is that virtually all recent work can be located somewhere upon a spectrum from the more internalist [e.g. 55; 63] shading through to the more externalist [e.g. 161; 125]. But the new eclectics, unlike externalists, recognize that scientific judgements about pertinent experimental or analytical results, or about correct theory, can sometimes only be understood in terms of the technical tradition within which they play a part, and may be insulated from wider social

6

considerations. This is not tantamount to internalism, however, since eclectic historians of science would argue (or assume) that in such cases the technical tradition itself is a socially constructed, or culturally determined, phenomenon and that work within that tradition is affected by social interactions between the relevant specialists [199: 352–3; 193: 222; 195; 75].

The important thing to note about the historiography of science, then, is that ever richer contextualization has been the main ambition of the majority of its practitioners for a number of decades. The result is a subdiscipline of history which is flourishing in its own terms, and which more generally is making a major contribution to our understanding of how and why science has become such an overwhelming feature of Western culture. Within this general effort to understand the cultural dominance of science, accounts of the Scientific Revolution play a major role. But it is important, whenever we consider these accounts, to bear in mind the complex historiographical issues which have helped to shape them.

2 The Scientific Method

The development and establishment of the characteristic methodology of science has always been regarded as constitutive of the Scientific Revolution. A. C. Crombie tried to suggest that at least one major element of that methodology, experimentalism, could be traced back to the thirteenth century [41; 42], but the tide of historiographical consensus is definitely running against him [139: 361]. Similarly, although there were mathematical sciences throughout the medieval period, it is generally acknowledged that the period of the Scientific Revolution saw a dramatic change in conceptions of, and attitudes towards, the mathematical analysis of nature. In this chapter we will consider both of these major components of the scientific method in turn.

(i) The mathematization of the world picture

The 'mathematization of nature' which has been seen as an important element in the Scientific Revolution used to be attributed to a sea-change in the metaphysical system which underwrote all concepts of the physical world, introducing 'Platonic' or 'Pythagorean' ways of looking at the world to replace the Aristotelian metaphysics of medieval natural philosophy. Recent work has shown the inadequacy of this view on a number of grounds and has pointed to an alternative account of changing attitudes to mathematics [107; 12; 238]. To put it simply, the Scientific Revolution saw the replacement of a predominantly instrumentalist attitude to mathematical analysis with a more realist outlook. Instrumentalists believed that mathematically derived theories are put forward merely hypothetically, in order to facilitate mathematical calculations and predictions. Realists, by contrast, insisted that mathematical analysis reveals how things must be; if the calculations work, it must be because the proposed theory is true, or very nearly so [107: 140].

The new realism can be seen at work in the astronomy of Nicolaus Copernicus (1473–1543). Astronomy, one of the so-called 'mixed' sciences, had always consisted of a mathematical and a physical part. Essentially, what this meant was that the astronomer had to reconcile the putative mathematical structures, which provided him with his means of calculating planetary and other heavenly movements, with the demands of Aristotelian cosmology and physics. Although the great synthesizer of ancient Greek mathematical astronomy, Claudius Ptolemy (AD 90–168), was a realist, his astronomical system had been regarded increasingly throughout the Middle Ages as a hypothetical system which, while providing a basis for calculation, was incompatible with the Aristotelian system. The result was a separation of the mathematical and physical parts of astronomy [238].

Where Aristotle's cosmology was a neat homocentric nesting of heavenly spheres in which only uniform circular motion could take place, Ptolemaic mathematical astronomy had planets moving on an epicycle, whose centre inscribes a circle (the deferent) in the body of the planetary sphere, in order to account, partially, for variations in the speed and brightness of the planet, and for its retrograde motion (a period in which the planet moves in the opposite direction to its usual progress). In spite of these devices, the required fit with observations could not be achieved without also assuming that the epicycle moved with uniform motion only with respect to an eccentric point, not with respect to the centre of the deferent, or with respect to the earth. Although such uniform motions around what was called the 'equant' point could be easily defined mathematically, it was by no means clear what kind of physical mechanism could explain such motions. Indeed, it was axiomatic in Aristotelian physics that all heavenly motions were natural, unforced, motions, and that the natural tendency of heavenly bodies was to move uniformly, in perfect circles [135; 43; 60; 168].

Ptolemaic astronomy was also beset with more pragmatic difficulties. Perhaps the most embarrassingly visible of these, by the end of the fifteenth century, was its inability to accurately set a date for Easter. Copernicus was concerned to solve this and other practical problems but he went much further by proposing his new system of astronomy in which the sun replaced the earth as the central body around which the planets, now including the earth, revolved.

There has been a tendency to see Copernicus, not as a truly revolutionary figure in the history of science, but rather as an essentially conservative thinker. For Thomas Kuhn, for example, Copernicus wrote 'a revolution-making rather than a revolutionary text' [135: 135; see also 32: 123–5]. Certainly, for those who like to draw up inventories, Copernicus is easily made to look conservative. He kept his book back for thirty years or so before being persuaded to publish. He made few astronomical observations (he was no revolutionary advocate of empiricism); he did not commit himself on the status of the heavenly spheres (he dithered as to whether they were solid, crystalline spheres in which the planets were embedded, or mere geometrical constructs [but see 238: 112–16]); he continued to believe in a finite sphere of fixed stars, even though his theory demanded that it was much larger than before [43: 99–0]. Furthermore, he refused to employ Ptolemy's equant point on the highly conservative grounds that it violated the Ancient precept that the heavenly motions must be uniform and perfectly circular, but otherwise he used Ptolemy's mathematical techniques (eccentrics and epicycles), to the extent that he even used epicycles upon epicycles to solve some problems [60; 135]. Furthermore, he looked backwards to Ancient thinkers to find precedents for the theory of a moving earth, finding them in various Pythagorean writers [135].

Even so, as Robert S. Westman has shown, Copernicus must still be regarded as a radical innovator in astronomy, and in the formation of a new role for the astronomer – as a natural philosopher [238]. For, while Andreas Osiander (1498–1552), a Lutheran pastor who saw Copernicus's *De Revolutionibus* through the press, took it upon himself (without permission) to add an anonymous preface in which he denied the ability of the astronomer to reach true conclusions about the actual nature of the heavens, Copernicus made very clear in his *own* Preface that he thoroughly repudiated the instrumentalist approach. Because his heliostatic system accounted for all celestial observations as accurately as Ptolemy's, while disposing of the unexplained annual component in each planet's movement (which is, of course, a transference of the earth's motion), and providing an easy and certain means of determining the order of the planets (arbitrary in Ptolemy), and their distances from the sun, Copernicus believed that his system must be taken to be physically true. So, Copernicus not only put the earth in motion against all the teachings of Aristotelian physics, the Holy Scriptures and common sense, but he also did so on what most contemporaries would have regarded as illegitimate grounds. No matter how contrary to natural philosophy the motion of the earth may seem, Copernicus insisted, it must be true _because the mathematics demands it._ This was revolutionary.

Immediately, we have to ask why Copernicus should have made such a bold step. Clearly, we cannot answer this question by reference to a unique cause, but we can point to a number of determining factors, all of which must be weighed up in our conclusions. Technical issues cannot be overlooked but nor can they provide the whole answer. No matter how good an astronomer and mathematician Copernicus was, he might have remained content to present his theory hypothetically, as Osiander pretended. It seems clear that Copernicus should be seen as a participant in a wider trend among mathematical practitioners, and perhaps a handful of humanist sympathizers, to improve their intellectual and social status [238; 12].

The factors involved in stimulating this trend were varied and complex but they include the recovery of Ancient Greek mathematical texts by humanist scholars, which provided new resources for making claims about the unity of mathematics, its usefulness and its certainty as a means of establishing truth [12]. There was also a successive weakening of the dominant Aristotelian natural philosophy which invited alternative views not only from professional natural philosophers (within the universities) but also from practising mathematicians, physicians and others. Changes in the court structure in Renaissance Europe also played a major part in enabling the elevation of at least some mathematical practitioners [12; 13; 64; 137], enabling them to escape from the restrictions placed upon them within the university system, for example, where there was a strict hierarchy of disciplines and subdisciplines in which the mixed mathematical sciences came below physics.

11

Even so, it would be wrong to see Copernicus as being carried along by such changes. By insisting upon the physical truth of his theory when his grounds for doing so were entirely mathematical he was contributing in a major way to the eventual triumph of this still insecure movement, not simply performing unremarkably within it. This is made clear when we consider that most astronomers used Copernicus's system only as a way of calculating planetary positions. Robert S. Westman's European-wide survey has led him to conclude that only ten thinkers accepted the physical truth of Copernicus's theory before 1600 [238: 136]. Interestingly, of these, only two worked all their lives as academics within the university system, and they were both Lutheran Germans affected by important pedagogical reforms introduced by the leading Lutheran theologian, Philip Melanchthon (1497–1560) [238: 120–1].

The subsequent history of the mathematization of the world picture shows the same crucial theme recurring. The major innovators are all concerned with the epistemological status of mathematics. Consider, for example, the Danish astronomer Tycho Brahe (1546–1601), arguably the best observational astronomer of his age. Tycho, like Copernicus, was free of the constraints of the university disciplinary hierarchy, and as a nobleman in his own right he was even free, unlike Copernicus, of the need to win patronage and support. Rejecting the motion of the earth he developed instead a compromise position between Copernicus and Ptolemy in which all the planets circled the sun, while it circled around a stationary earth. It is obvious enough from this that Tycho was a mathematical realist, but his work first generated controversy amongst natural philosophers when he published the results of his observations of a new star in 1573 (evidently what we now call a supernova) and of comets in 1588 [43: 137–46; 215]. The new star caused problems for traditional Aristotelianism in so far as the heavens were supposed to be perfect and not subject to change. These problems were compounded when he was able to establish that comets, observed in 1577, 1580 and 1585, were superlunary phenomena, one comet being at least six times further from the earth than the moon. Previously, in accordance with Aristotelian doctrine, comets and meteors were held to be atmospheric phenomena (which, by the way, is why our word for the study of the weather is still meteorology), but Tycho demonstrated that they were not. Encroaching still further upon the territory of natural philosophy, Tycho showed that the comet's path took it through the heavens in such a way that it must shatter the doctrine of crystalline spheres. Henceforth, the planets had to fly through space independently. This was to have profound implications. It is one thing to envisage heavenly motions in terms of spheres rotating about their axes – motions without any change of place (remember: the 'sphere of Mars', say, refers to a sphere completely surrounding the centre of the world system, which rotates upon an axis and carries the visible marker, Mars, around with it). It is quite another to have to consider planets as independent bodies, actually moving across vast distances of space. While it might be said that it is the nature of a sphere to revolve about itself without need of a driving force, the continuous motions of freely movable planets seem to require something more in the way of an explanation.

Johannes Kepler (1571–1630), undoubtedly the greatest Copernican astronomer, was himself so concerned with the right of the astronomer to be considered as a natural philosopher that he made it the major theme of a formal defence of Tycho (in a priority dispute) that he was obliged to write in order to secure Tycho's patronage [126]. Furthermore, it seems true to say that Kepler would never have made his own contributions to astronomy if it had not been for his mathematical realism and his conviction that the astronomer must also be a natural philosopher. In his *Astronomia Nova* of 1609 he not only discovered that the planets take elliptical paths around the sun (which is situated at one focus of the ellipse), and that the speed of the planet varied continuously, increasing as the planet approached nearer to the sun, decreasing as it receded, but he also proposed a physical explanation for those movements. Indeed, the full title of his *New Astronomy* indicated that it was 'based on a theory of causes' to provide a 'celestial physics' (*Astronomia Nova Aitiologetos, Seu Physica Coelestis*). Taking inspiration partly from the magnetic philosophy of William Gilbert (1540–1603), and partly from the Neoplatonic tradition of light metaphysics (closely linked to the mathematical tradition of geometrical optics), Kepler suggested that the planets, including the earth, might have something analogous to a magnetic axis which kept them continually orientated the same way in space, and which could produce alternate phases of attraction towards and repulsion from the sun (if we imagine the sun as a magnetic monopole). Kepler insisted that magnetism should be seen only as an example of the kind of force that might be involved, however, and he also invoked the light of the sun as another analogue of the kind of thing he had in mind [134: 185–224]. Recent research has revealed that Kepler finally became convinced of the truth of elliptical paths because they were amenable to this kind of physical explanation. He was able to find a number of alternative geometrical schemas which would account equally well for Tycho's observations, but his search for physical causes to account for the geometry led him to what are now known as his two first laws of planetary motion [209; 71; 134; 41].

One important aspect of changes in astronomical theories from Copernicus to Kepler was the increasing realization that the strict Aristotelian division between sublunary phenomena and superlunary, or heavenly, phenomena (in which, for example, all natural motions below the sphere of the moon were rectilinear motions, while natural motions in the heavens were always circular) was no longer tenable. By removing the earth from the centre of the universe, Copernicus compromised the notions of 'up' and 'down' which essentially defined sublunar natural motions. Furthermore, the dissolution of the heavenly spheres removed the standard mechanism for explaining the motion of the planets (usually held to be caused by a driving force frictionally transmitted from one crystalline sphere to the next [138]) and demanded a new account. This challenge was taken up not only by Kepler, but also by William Gilbert, Galileo [205], Giovanni Borelli (1608–79) [134], Isaac Beeckman (1588-1637), Descartes and numerous others [2; 233], until Isaac Newton's *Mathematical Principles of Natural Philosophy* won general acceptance as the correct solution [233; 236; 100].

But the new use of mathematics to explain, not just to describe, the workings of the physical world was not confined to celestial matters. The growth of trade, the beginnings of colonization and the concomitant drive to exploration meant that practical mathematical techniques like navigation, surveying, and cartography came to be seen as much more important, attracting the interest of some leading intellectuals and enabling some lowly practitioners to raise their social and intellectual status [9; 10; 40]. The mathematical science of (terrestrial) mechanics, which could be subdivided into statics, hydrostatics, and kinematics, also saw remarkable changes during our period. Once again, to understand these changes we must consider technical developments together with significant shifts in the social role of mathematicians. Innovations in warfare, and in particular the ingenious response to cannon siege, the artillery-resistant bastion, and various civil engineering schemes such as land reclamation, canal building, or even just surveying for fiscal purposes, have been seen as major causes not only of the increased status of mathematicians in early modern Europe, but also of the increased interest in mathematics shown by members of the patrician class [12; 137; 131].

Changes in the nature and structure of royal courts in a Europe of increasingly absolutist states also expanded the opportunities for the *mathematicus* to make his presence felt. The mathematical practitioner who could impress the prince by his production of *mirabilia*, striking machines or sets for masques, and other enhancements of the prince's image, could rise above those involved merely with the management of the estate. These mathematicians, because of their position at court, could easily flout the hierarchical distinction between mathematicians and natural philosophers that existed within the university system. So, Giovanni Battista Benedetti (1530–90) could leave his post as court *mathematicus* to Duke Ottavio Farnese in Parma to become Philosopher to the Duke of Savoy at Turin. Similarly, while Galileo remained a university professor, he was a lowly paid *mathematicus*, expected to defer to the higher status natural philosophers, but when he negotiated his position at the court of Cosimo de Medici, he could ask for, and be granted, the title of philosopher [12; 13; cf. 156].

Renaissance humanist discoveries and editions of the works of Archimedes, Pappus, Hero of Alexandria, and of a series of mechanical questions which were erroneously attributed to Aristotle, also stimulated, or made possible, a bolder attitude on behalf of those with the mathematical expertise to put these works to practical use. Increasingly throughout the sixteenth

15

century we see mathematicians dealing with terrestrial mechanics who were not content to present their work as merely descriptive, or as subservient to a traditional Aristotelian natural philosophy which was looking successively weaker. In the work of men like Simon Stevin (1548–1620) in the Netherlands and Nicolo Tartaglia (1500–57) in Italy [55; 131; 47], the separation between theory and practice, imposed by university professors of natural philosophy, was repeatedly exposed as untenable.

Of course the greatest figure in this movement is Galileo Galilei whose move from frustrated *mathematicus* in the university system to natural philosopher at the court of Cosimo de Medici is now seen to have been driven by the nature of his scientific ambitions and, in turn, to have had a visible effect on the content of his scientific work [13]. Although Galileo is most famous for his defence of Copernican theory, his initial interest was in terrestrial mechanics, and in particular kinematics. Like many of his contemporaries, he was dissatisfied with the Aristotelian account of motion, and struggled to arrive at a better theory. During the course of his career his account of free fall, for example, took him from a mere refinement of Aristotle's belief that bodies fall with speeds proportional to their weight, to the realization that acceleration in free fall is a constant (in a vacuum) for all bodies. He was also able to prove the parabolic path of projectiles, by assuming, contrary to Aristotle, that the natural motion of a body (its free fall) took place regardless of the forced, or unnatural, motions to which it was subjected. For Aristotle a projectile travelled in a straight line in the direction it was thrown, until the cause of its unnatural motion ceased, only then would it travel to earth by a straight line downwards under the natural motion of fall. Galileo's parabolic path was shown to derive from a combination of these two motions (natural and unnatural) acting simultaneously [59; 204; 99: Ch. 4]. The claim that two motions could take place simultaneously also helped Galileo to answer various objections to the Copernican theory. A ball dropped vertically from a tower, he insisted, would not fall far to the west of the tower as the rotating earth moved eastward during the ball's descent. The descending ball would stay close to the tower because it already had the same component of circular motion as the tower and everything else on the earth.

16

Another of Galileo's theories of motion was developed as a means of trying to account for the motion of the earth around the sun. As a confirmed Copernican who wished to prove himself as a natural philosopher, Galileo took it upon himself to explain how a body like the earth, weighing countless tons, could be kept in perpetual motion. This was a major difficulty for Copernican theory. According to Aristotle, everything which moves is moved by something. So what pushes the earth around? Kepler, as we've seen, tried to derive a motive force by analogy with magnetism and light; Galileo simply denied the Aristotelian assumption that motion required a continuous cause. In a brilliant passage of his *Dialogue on the Two Chief World Systems* (1632) Galileo argued that, whereas a smooth ball on a frictionless inclined plane will accelerate continually as it moves down the slope, and will rapidly slow down if it is made to move up the slope, such a ball will have no tendency either to speed up or slow down when the plane is perfectly horizontal. Once set in motion on a horizontal plane, therefore, the ball should continue to roll indefinitely at the same speed. But a horizontal plane in this context, means one in which all its parts maintain an equal distance from the centre of the earth, which would in fact be a sphere all around the earth if it was extended. Accordingly, he was able to suppose that just as a bronze ball might move perpetually around the earth in a perfect circle, so might the earth itself move perpetually round the sun. Clearly, this argument would be completely undermined by any suggestion that planets did not move in perfect circles, but rather in ellipses, in which they did indeed approach to and recede from the sun, and so it is hardly surprising that Galileo never seems to have paid any serious attention to Kepler's astronomical conclusions. The *Two Chief World Systems*, mentioned in the title of his great book, were the Ptolemaic and the Copernican.

Galileo excluded the Tychonic compromise system as well, even though it was perfectly compatible with the various astronomical discoveries that Galileo had made earlier with the newly invented telescope. In his *Siderius Nuncius* (*Starry Messenger*) of 1610, Galileo presented evidence which enabled him to claim the moon was just like the earth in composition (with mountains, valleys, seas, etc.), and not a qualitatively different body composed of an unearthly fifth element (or quintessence) whose natural motion was to move in a circle. The implication was clear: if the moon could move around the earth, even though it was made up of tons of earth and water, why couldn't the earth itself move around the sun? The phenomenon of earthshine, which he saw faintly illuminating the dark side of the moon, showed that the earth did not differ from the planets by its lack of light. The moons of Jupiter which he discovered suggested that it was possible for each planet, not just the earth, to have its own moon(s) and be able to move through space without losing them. And the discovery of countless stars invisible to the naked eye gave credence to the post-Copernican suggestion that the fixed stars were not confined to a sphere but were spread out through infinite space. These discoveries and the later discovery of sun-spots were irremediably damaging to Aristotelian precepts but neither these nor the extra discovery that Venus showed phases like the moon, did any damage to the Tychonic system.

Galileo was undoubtedly a versatile and creative thinker but research has shown the importance to him of the work of his predecessors, whether they be older contemporaries among the *mathematici*, men like Tartaglia or Guidobaldo del Monte (1545–1607), medieval thinkers like those who developed impetus theories to account for projectile motion, or professors at the Jesuit Collegio Romano [224]. It is also known that there was nothing new in his *modus operandi* which was essentially that of other *mathematici*, combining mathematical analysis and experimental investigation [205; cf. 10]. Nevertheless, Galileo was a forceful publicist of his own ideas and a superb communicator of technical ideas. His major works were published in Italian, rather than the Latin of the scholars, and were quickly translated into other European languages. Perhaps his greatest contribution to the development of science was, as Gary Hatfield has recently argued, his exemplification of the usefulness and success of the mathematical approach to nature [107: 139]. Repeatedly in his writings, Galileo teaches by example, showing how mathematical practice can help us to understand the nature of the world; even in those cases where the fit between mathematical analysis and physical reality is only approximate, the mathematics being based on idealized, and unrealizable, circumstance.

Another important contribution to the mathematization of natural philosophy was provided by the Jesuits with their vigorous pedagogical activities. Mathematics played an important role in their so-called *Ratio studiorum* (Order of studies). The Jesuits signalled the importance they set upon mathematics by teaching it alongside physics, or metaphysics, in either the penultimate or final year of their course of study (rather than as a preliminary subject, taught at a very low level) [109: 101–14; 50; 138: Ch. 2]. The significance of Jesuit pedagogy can hardly be overstated and the attitude to mathematics propagated in their colleges, while undoubtedly an expression of the general trend we have been describing, cannot have failed to reinforce the importance of mathematics for understanding the world in the minds of the students.

There are at least two famous students of the Jesuits who made their own influential contributions to the mathematization of the world picture: Marin Mersenne (1588–1648) and René Descartes. Marin Mersenne became a friar of the Order of Minims in 1611 and, with the encouragement of his Order, spent his life in intellectual labour in support of his faith. Mersenne was led by his religious beliefs to deny the fundamental assumption of Aristotelianism that physical causes could be known with certainty. This was to claim that humankind was capable of penetrating to the essence of a thing, and so equal to God. Mersenne was no sceptic, however. He saw mathematics as the most certain kind of knowledge; the only form of human knowledge which could aspire to be the equal of divine knowledge [47]. Mersenne played an active role in publishing his ideas and those of other mathematicians, but he was even more active in cultivating and maintaining an extensive correspondence with leading intellectuals all over Europe. Inevitably he sought out like-minded individuals and acted as a major source of information for each of them, communicating current work to interested parties. In so doing, of course, he could hardly fail to communicate his own ideals and his own fundamental belief in the importance of mathematics to philosophy.

Descartes is now known primarily as a philosopher, but at the outset of his career he was a mathematician, working on music, optics, and mechanics. Indeed, his famous *Discourse on*

19

Method of 1637 (in which he put forward that most famous of all philosophical arguments, *Cogito, ergo sum* – I think, therefore I exist) was published as a preface to three exercises in mathematical physics (on the sine law of refraction, the cause of the rainbow, and how to represent abstract algebraic problems in spatial or geometrical terms) which were supposed to exemplify the power and certainty of that method [88; 89; 207; 83]. Descartes' method led him to a new metaphysics, which provided the basis for a new system of physics, which in turn became the most influential of the new 'mechanical' philosophies (see Chapter 4). Although his final system made less use of mathematics, being rather more speculative and qualitative, there can be no doubt that it grew out of Descartes' early concerns to understand the physical world in mathematical terms.

The cast of characters who played an important role in the mathematization of the natural philosophy could easily be extended. Galileo effectively founded a school of followers who carried on his work in mathematical physics; men like Bonaventura Cavalieri (1598–1647), Evangelista Torricelli (1608–47), and Borelli [194]. The Low Countries provided fertile ground for higher mathematics [36]. Isaac Beeckman set an impressive example of how to use mathematics in natural philosophy. Although he published nothing, his work was known to others through Mersenne, or through personal acquaintance. He was a particularly important early influence upon Descartes [89]. Mathematical physics in the Netherlands reached its peak with Christiaan Huygens (1629–95), who is all too often presented as an important forerunner of Isaac Newton (because his development of the concept of centrifugal force made it clear that motion in a curved path, like that of the planets, requires the constant action of a force, so paving the way for the Newtonian affirmation of rectilinear inertia and the rejection of lingering ideas, deriving ultimately from Galileo, that motion in a circle can also continue indefinitely). But Huygens can be seen, less whiggishly, as one who first of all adapted and refined the mechanistic philosophy of Descartes which he had found wanting on both mathematical and methodological grounds, and who subsequently developed a mechanistic philosophy in opposition to what he saw as the non-mechanical philosophy of Newton [244; 233].

Isaac Newton's *Mathematical Principles of Natural Philosophy* (1687) can be seen as the culminating point of the mathematization of the world picture. Most famous for establishing that the planets continue to orbit the sun as a result of the same force which makes an apple fall to the ground, the *Principia Mathematica* did much more besides. It demonstrated mathematically the truth of Kepler's laws of planetary motion and initiated modern lunar and cometary theory. It showed the usefulness of mathematics to an understanding of both the celestial and terrestrial realms, and finally refuted the Aristotelian distinction between sublunary and superlunary physics. Newton's laws of motion displaced Descartes' laws and formed the basis of a complete understanding of the behaviour of colliding bodies (including oblique collisions, which had completely defeated Descartes). Newton was able to deal fully with centrifugal force, and to make a beginning towards understanding the motions of bodies in resisting fluids. The latter enabled him to develop a theory of acoustics in which the velocity of sound varied with the pressure and density of the medium through which it passed. Of crucial importance for the mechanical philosophy, to which he and the majority of his contemporaries subscribed, he demonstrated mathematically how observable macroscopic effects could be explained in terms of microscopic phenomena [79; 100; 233; 236].

The publication of Newton's *Principia* marks the completion of the trend towards the mathematization of natural philosophy which began in the sixteenth century. But perhaps it is true to say that we make that judgement about the *Principia* because Newton, unlike Galileo or Descartes, succeeded in getting the mathematics and the physics substantially correct. Newton himself did not have to justify the mathematical approach; he could safely assume that there was an audience for his book, who, even if they could not follow its mathematics, took for granted the validity of mathematics for understanding the workings of the world [80]. Although his book met with some fierce criticism, not a murmur was raised against it in this regard. That battle had already been won, and in a sense the story was already complete before Newton stepped onto the stage. Certainly, by the end of the seventeenth century the mathematician was regarded, not as a mere underworker to the natural philosopher, but as one of the intellectual elite.

21

This fact is strikingly illustrated by the short shrift given to Robert Hooke (1635–1703) concerning his claim that Newton had the main principle of celestial mechanics from him. It's certainly true that, in an exchange of letters in 1679, Hooke told Newton that he could account for Kepler's laws of planetary motion on the assumption of a single attractive force towards the sun operating on the tangentially moving planet. He even indicated that the force must be taken to vary inversely as the square of the distance between sun and planet. It is also now known that prior to this exchange Newton had been trying to describe celestial motions, in typical Cartesian fashion, in terms of a balance between two forces: a centrifugal force caused by revolution about the centre, and the centripetal force of gravity. Newton adopted Hooke's assumptions in the *Principia* [236: 382–8] but when Hooke called for some acknowledgement that he had provided Newton with the idea, Newton took exception to the implication that mathematicians are 'nothing but dry calculators & drudges' working to service the man of ideas. Hooke found one or two who would take his part among his circle of friends, but most then, and all since, seem to have dismissed his idea as trivial compared to Newton's working out of the precise mathematics involved. 'The discovery was Newton's', the historian R. S. Westfall has insisted, 'and no informed person seriously questions it' [236: 448–52]. The message seems to be clear: the real natural philosopher is also a mathematician. In the treatment of Hooke, then, we see an early example of the kind of awestruck attitude to mathematical physicists that still pertains today. This in itself is a major legacy of this aspect of the Scientific Revolution.

Our aim here is merely to account for the Scientific Revolution and we have seen how the rise of the mathematical approach to our understanding of the natural world from the sixteenth through to the end of the seventeenth century distinguishes the period from what went before, and resulted in some dramatic changes in the conceptualization of physics. This is by no means an adequate history of early modern physical science, much less of mathematics, however. There are a number of important elements we have hardly mentioned. A similar story might have been told by concentrating more on the development of geometrical optics [177; 183] or even on theories

of music and harmony (a concern, to a greater or lesser extent, of Galileo, Kepler, Beeckman, Descartes, Mersenne, Hooke,
Huygens, and Newton, as well as others we have not discussed)
[34; 30; 58: Ch. 2; 59: Ch. 1; 210]. And there are numerous
names we might have invoked: from Gemma Frisius (1508–55)
to Blaise Pascal (1623–62), from Egnazio Danti (1536–86) to
Pierre de Fermat (1601–65), from Roberval (1602–75) to Leibniz
(1646–1716). We have overlooked numerous topics: the principle of conservation of a vector, conservation of momentum,
the mathematics of indivisibles and the development of calculus. Each of these, individuals or topics, and many more, contributed importantly to the Scientific Revolution, and a close
study of any of them would provide further support not only
for the role of the mathematization of nature in the Scientific
Revolution, but also for the importance of the social context
in understanding this process.

(ii) The experimental method

The rise of mathematics followed the rise of mathematicians:
the mathematical approach to the understanding of nature grew
more persuasive as the mathematician became more authoritative. The mathematician began to acquire the cognitive authority previously reserved for the natural philosopher. One way
in which this new authoritativeness was acquired by mathematicians was by laying claim to the certainty of mathematical knowledge [127; 12]. But such claims were easily disputed, particularly
if mathematics was being used to claim something as implausi-
ble as, say, the motion of the earth. After all, mathematics was
an artificially constructed system, and the certainty of its claims
was conditional: if you accept certain axioms and other precepts, then you have to accept the various conclusions which
can be demonstrated to follow from them. But why should the
axioms and other preliminaries relate to the physical world
in any way? For example, how can the suggestion that one
negative amount multiplied by another negative amount always
yields a positive amount have any relevance to the way things
really are?

Furthermore, within the dominant scholastic-Aristotelian tradition the authoritative claims of natural philosophy were based upon what were held to be the evident, undeniable, truths of experience. Mathematical claims, however, have a striking tendency to be far from evident. Thomas Hobbes (1588–1679), famously, became fascinated by Euclidian geometry upon seeing one of Euclid's theorems for the first time and being struck, not by its obvious truth, but by its seeming impossibility. In order for mathematicians to establish the validity of their approach to understanding the world, they had to establish new criteria of assent, new principles of authority [50].

Mathematical practitioners, therefore, became important contributors to the new trend towards experimentalism. For one of the characterizing features of the Scientific Revolution is the replacement of the self-evident 'experience' which formed the basis of scholastic natural philosophy with a notion of knowledge demonstrated by experiments specifically designed for the purpose. Like a mathematical proof, the end result of the experiment might well be knowledge which is counter-intuitive.

Unfortunately, the precise nature of the role of the mathematical sciences in the formulation of the experimental method has not yet been established by historical research, although there are a number of highly suggestive studies [9; 10; 50; 59; 138; 205]. It seems fairly clear, nonetheless, that mathematical practitioners played an important part in the establishment of the experimental method.

One way of seeing the connection between mathematics and experimentalism is through the history of scientific instruments. Before the Scientific Revolution the only instruments in use were armillary spheres, astrolabes, quadrants, and one or two other instruments used exclusively by astronomers. In the sixteenth and seventeenth centuries, however, a much more diverse range of mathematical instruments came into use to facilitate problem-solving in all branches of the mathematical disciplines. But this same period also saw the development of the first natural philosophical instruments; that is to say, instruments intended to discover new truths about the nature of the world. Principal among these were the telescope and microscope, the barometer and air-pump, the thermometer, and later various electrical machines [219; 9; 104]. It is surely significant that the telescope, a new invention used commercially, first became an instrument of natural philosophy in the hands of Galileo, a mathematical practitioner with a burning ambition to be acknowledged as a natural philosopher. The Galilean telescope can be seen as an extension of those earlier astronomical instruments which enabled Tycho Brahe to establish that the new star *was* a new star and that comets too were heavenly, not atmospheric, phenomena. The telescope became a scientific instrument as it became integrated into the purposes of astronomers, and in so doing it made it easier for astronomers to become natural philosophers, pronouncing upon the real nature of the heavens [220]. It is reasonable to assume, therefore, that it was the mathematical tradition which first provided the stimulus towards the use of instruments in scientific research. As J. A. Bennett has suggested, there seems to be a clear progression from the mathematical instrument as the trademark of the mathematical practitioner to the scientific instrument as the hallmark of the modern scientist [9; cf. 220].

The mathematical sciences were always concerned with practical, useful knowledge and the practitioners were generally empiricist in their orientation, testing the application of their mathematical techniques to the real world. This is most clearly seen, perhaps, in attempts to use the discovery of magnetic variation as a means of determining one's longitude at sea. A ship's position north or south of the equator was easy to determine by reference to the sun or stars, but determination of its position east or west of a given reference point had defeated all efforts. When it was realized that a compass needle orientates itself not to the geographical north pole but to a fixed point (later discovered to vary in position) some distance away from the pole, it seemed to provide a possible solution to the problem. In all these efforts it was necessary to test calculations against the observations of mariners, and in a number of significant cases empirical studies of magnets and their behaviour also played an important part in the investigations [9; 245; 225]. It seems as though mathematical practitioners undertook this kind of empirical testing of their work almost routinely. In view of this, it looks very much as though the mathematical tradition must be regarded, as Bennett has urged, as a major source of the experimental method in seventeenth-century science.

25

If the Renaissance expansion of overseas exploration, trade and colonization demanded improvements in navigation and other mathematical aspects of geography, it was not the only sphere where skilled craftsmen could make contributions that university-trained intellectuals could not. The increased importance of mining and metallurgy in the economy of sixteenth-century Europe led to increased interest in such matters from men of higher intellectual status. This interest is manifested in two influential books, Vanoccio Biringuccio's (1480–*c.*1539) *De la Pirotechnia* (1540) and George Agricola's (1490–1555) *De Re Metallica* (1555). These works clearly illustrated the relevance of craft knowledge to an understanding of the nature of the world, and reinforced the teachings of humanist pedagogues like Juan Luis Vives (1492–1540) who advocated the study of trade and craft secrets. They also emphasized the importance of experience in the foundation of knowledge. Increasing awareness of the more practical knowledge of elite craftsmen has been seen as a major factor in the development of the experimental method. Bernard Palissy (1510–90), a much admired potter who tried to discover for himself, by trial and error, the secret of Chinese enamelware, gave public lectures in Paris on agriculture, mineralogy and geology, and published *Discours admirables* (1580), dialogues in which the superiority of 'Practice' was extolled over blinkered 'Theory' [180].

The mathematical tradition was not the only historical source for an experimental approach to the understanding of nature. There were also a number of significant developments in anatomy and physiology. For example, a revolutionary break with previous ways of teaching anatomy in medical schools occurred in the University of Padua, with the appointment in 1537 of a humanist-trained scholar who also happened to be a very adroit dissector, Andreas Vesalius (1514–64) [76; 162]. Vesalius taught anatomy while doing his own dissections (it was more usual for the lecturer to read from the ancient authority, Galen, while a surgeon did the actual dissection), and proved immensely popular with medical students. Furthermore, his great book, *De Humani Corporis Fabrica* of 1543, was both a textbook of anatomy and a superbly illustrated practical manual on how to dissect. And Vesalius took care to provide a preface in which

26

he deplored the separation of surgery (at that time a craft tradition) from medicine. As a result Vesalian anatomy came to be seen, by some at least, as 'the foundation of all medicine' and threatened for a while to supplant natural philosophy from its position at the centre of medical education [226; 22].

Vesalius allegedly discovered 200 errors in Galen's anatomical writings, and the most important of these, his discovery that the wall within the heart, separating the right ventricle from the left, was not perforated, threatened the whole of Galenic physiology [162]. Although Vesalius himself did not go any further with his studies of the heart, his successors at Padua made a number of pertinent discoveries. Realdus Columbus (1510–59) put forward the theory of pulmonary circulation (in which blood traversed from the right ventricle of the heart to the left by crossing the lungs, rather than by seeping through putative perforations in the intervening muscular wall), while Hieronymus Fabricius (1533–1619) discovered the valves in the major veins of the leg, which William Harvey (1578–1657) later realized allowed a flow of blood only towards the heart.

Harvey was at Padua from 1600 to 1602 and was schooled by Fabricius in what Andrew Cunningham has called the 'Aristotle project' – a concerted effort to acquire true causal knowledge of the parts, or organs, of animals and of their generation, and so to raise concern with living things from the status of natural history (descriptive) to that of natural philosophy (prescriptive) [44; cf. 77]. On his return to England he carried on this Paduan tradition by studying the generation of animals, and the motion of the heart and blood. Research into the latter led him to the discovery of the circulation of the blood. In his *De Motu Cordis et Sanguinis* (1628) Harvey not only devised ingenious but uncomplicated experimental techniques, but he also drew upon the craft knowledge of butchers and slaughtermen (he was able to explain why they sever an artery to drain all the blood from the body, and why, if this is not done, the arteries will be empty after death but the veins will be full of blood).

Careful research now enables us to see that Harvey was not 'ahead of his time', a modern thinker in seventeenth-century dress. While most contemporary anatomists confined their studies to man, the Paduan 'Aristotle project' was wider in scope, seeking to understand the form and function of parts more generally in animal systems. Accordingly, Harvey investigated the motions of the heart and blood in animals other than man. This meant that he could perform vivisection experiments – something never considered by most anatomists. Many of Harvey's discoveries were made because he experimented like a Paduan Aristotelian. Moreover, his vitalism ensured that he thought in a way that was very different from a modern biologist [45]. He was convinced that the blood contained some principle 'which corresponds to the element of the stars', and which is the principle of life and the soul within it [165; 101]. Furthermore, although he changed his mind during his career about whether the heart or the blood itself was the prime seat of life, he never doubted but the heart-blood system was self-contained and needed no other principle, such as air from the lungs, to re-vivify it.

It can be taken as a sure sign of the authoritativeness of Harvey's experimental demonstrations that his theory was taken up by others remarkably quickly (although there was, of course, vigorous opposition from some quarters [240; 77]). This is all the more remarkable given that it completely undermined Galenic physiology without offering any new system to put in its place. We might expect Harvey's discovery to lead to a collapse of the Galenic system of therapeutics, which was so intimately bound up with the physiology, but it didn't. The Galenic system of therapeutics, because of its practical success in dealing with illness (bear in mind that doctors had been making a good living from it for centuries), remained remarkably untouched by Harveian innovations, although there were for a short time in the 1660s some dangerous experiments in London and Paris with blood transfusion and injections [75].

Physiology did, however, become the focus of major experimental investigations. The role of the lungs and respiration, of the liver, and the nervous system, all left unexplained by Harvey, were just some of the major foci of research. The Harveian experimental research programme was richest perhaps in England, flourishing from the late 1630s to the mid-70s, and involving major figures like George Ent (1604–89), Nathaniel Highmore (1613–85), Thomas Willis (1621–75), Christopher Wren (1632–1723), Robert Hooke, Robert Boyle, and Richard Lower (1631–91) [75]. But his influence stretched to the continent, shaping subsequent experimental research and teaching in the medical schools [77].

There were revolutionary developments in natural history too. Renaissance humanists like Otto Brunfels (c.1489–1534), Leonard Fuchs (1501–66) and Gaspard Bauhin (1541–1613) were concerned to extend the ancient encyclopaedic surveys of the plant and animal worlds provided by Aristotle, Theophrastus, Pliny and Dioscorides, to take into account species from northern Europe, or from the Americas, unknown to the Ancients. The difficulty of recognizing ancient descriptions made precise identification a major concern and the printing press was put to good advantage by illustrating the new catalogues. The importance of these uniformly illustrated texts cannot be exaggerated. They were a huge improvement on medieval manuscript herbals and bestiaries, in which illustrations, if any, were very unrealistic, either because they were copied (often very crudely) from an earlier version by a man who was trained as a scribe, not as a draftsman; or because their function was not to portray reality but the symbolic role of the creature in folklore (so, pelicans were shown feeding their young on their blood by deliberately wounding their own breasts). The attempted realism of the illustrations by skilled artisans, which unlike the more formalized decorative illustrations of earlier works invited comparison with real specimens, reinforced the explicit message of the texts that personal experience was a more reliable guide than authority, and the implicit message that skilled craftsmen had something to offer towards an understanding of the real world [6; 53; 72; 208].

If there was an increase in realism in the illustrations, there was a corresponding increase in what might be called the naturalism of the texts. The great Renaissance encyclopaedias of natural history, notably the four-volume *Historia Naturalium* (1551–8) of Conrad Gesner (1516–65) and the thirteen volumes on different kinds of animals published by Ulisse Aldrovandi (1522–1605), were concerned not just with mundane facts about the habits and the nature of the animals they discussed, but with the symbolic meanings these animals held for the ancients or for different contemporary peoples. As such these natural histories included in the entry for any given animal all the adages, proverbs, fables, Scriptural accounts and other folklore about the animal. It would seem that for natural historians like Gesner and Aldrovandi such information was relevant to an understanding of the animal itself, its nature and significance. Underlying this belief was a conviction that all creatures had myriads of hidden meanings and countless connections with other things, be they other animals, plants, minerals, heavenly bodies, numbers, or even man-made artefacts like coins, or amulets. Only by listing all that was known or said about the animal could all these putative connections be revealed. But this monumental kind of natural history, which seems to have clear connections with the traditional magical world-view that all things are connected in a Great Chain of Being and all have correspondences with other links of the chain, gave way in the seventeenth century to an entirely naturalistic kind of natural history. Confronted by animals and plants of the new world, which had no associations or similitudes for old world culture, no symbolic significance of any kind, natural historians produced encyclopaedias which presented all creatures, new world and old world alike, in more factual terms. To be sure, the culinary and medicinal uses of plants and animals might still appear as part of their natural history, but not their use in moral instruction [8; 72].

The social motivation towards this changing work in natural history was essentially two-fold. Firstly, it can be seen as an extension of Renaissance humanist concerns with the moral superiority of the *vita activa* over the contemplative life, embracing disciplines which were useful to the State, such as ethics, law, politics, and rhetoric, as well as a pragmatic, useful knowledge of nature. Knowledge of natural history was shown in these works to be useful in commerce, agriculture, cookery, medicine and a number of other areas which served the public good no less than the moral philosophy of civic humanism [37; 72; 212; 175]. An important aspect of the achievement of the leading propagandist for experimentalism, Francis Bacon (1561–1626), was that he codified and formalized the grounds for the philosophical authority of natural history by elevating inductive logic over deductive in his *New Organon* (1620) – a proposed replacement for the logic of Aristotle's *Organon*. The result was that naturalists like John Ray (1627–1705) could see themselves contributing to 'a philosophy solidly built upon a foundation of experiment' (1690) [175].

Secondly, natural history was seen as a way of displaying the marvellous wisdom, artistry and benevolence of the Creator. In this regard natural history could go far beyond the anthropocentric concerns of the *vita activa* and consider creatures which seemed to have no medicinal, culinary, or commercial value [6: 16; 53]. The result of this religious emphasis was that botanists and zoologists could lay claim to greater intellectual authority than was normally accorded to the merely descriptive discipline of natural history. The natural historian read God's second book, the book of Creation, to supplement the theologians' reading of Scripture.

The religious impulse towards natural history seems to have been dominant in the new studies using the simple or compound microscopes which were developed from about 1625. Jan Swammerdam (1637–80), a great Dutch microscopist and comparative anatomist, who showed by dissection that the wings of the future butterfly were present in the caterpillar (and so disposed of Aristotelian beliefs in the internal amorphousness of insects and their development by total metamorphosis), believed that the anatomy of a louse revealed 'the Almighty Finger of God' [37: 56]. Antoni van Leeuwenhoek (1632–1723), the discoverer of protozoa and bacteria, was also driven by physico-theological preoccupations [167]. It was possible, however, to use the microscope for more pragmatic concerns. While Swammerdam wrote a book on the mayfly and invested it with eschatological reminders, *Ephemeri Vita* (1675), Marcello Malpighi (1628–94) had earlier chosen to write on the more marketable silkworm, *Dissertatio de Bombyce* (1669).

Malpighi put the microscope to particularly good use in the study of the fine structure of anatomical features. He discovered the capillary connections between arteries and veins, and so set the seal upon the theory of blood circulation [1]. And yet by 1692, two years before his death, his fellow erstwhile microscopist, Robert Hooke, pointed out that only Leeuwenhoek continued to do serious work with the new instrument [241]. At least part of the reason for the failure of the microscope to become as essential to anatomical studies as the telescope became in astronomical studies was its inability to command authority among medical practitioners. The telescope's ability to increase the accuracy of positional astronomy guaranteed its usefulness, but knowledge of the invisible structure of organs did nothing to improve the efficacy of a medical system based essentially on the study and treatment of symptoms of disease. The microscope may well have been used by some for 'Diversion and Pastime', as Hooke suggested, but it needed to be taken up by practitioners in the relevant disciplines if it was to have a real impact. Instead, leading physicians like Thomas Sydenham (1624–89) and John Locke (1632–1704) explicitly rejected its use [242; 241].

Another major source of the experimental method is to be found in the chemical or alchemical tradition. Alchemy did not suddenly become experimental in the Scientific Revolution – it had always been an experimental pursuit. What did happen in the Scientific Revolution was that alchemical experimentalism began to make itself felt among natural philosophers, physicians and other intellectuals who were already becoming attuned to the teachings of experience through developments in the mathematical sciences, natural history, anatomy and medicine.

The major single influence was Paracelsianism, a philosophy which could take many forms but which always extolled the usefulness of practical chemistry in medicine and in a wider understanding of the natural world (the macrocosm) and of man (the microcosm). Founded by an itinerant Swiss autodidact who called himself Paracelsus (c.1493–1541), this chemical philosophy and new system of medicine became so influential that it was impossible to ignore [178; 53]. But Paracelsus was not just a vigorous propagandist for experientialism, his medical system was genuinely innovatory and, though it divided the ranks of medical practitioners, a number of therapeutic successes seemed to indicate it was a significant improvement on traditional medicine in at least some respects. As new chemically prepared medicines took their place alongside traditional herbal remedies in official pharmacopoeias all over Europe, the validity of empiricism looked more and more irrefutable [51; 54; 229].

Paracelsianism gained many adherents and a number of these – such as Gui de la Brosse (*fl.* 1630), founder of the Jardin des Plantes in Paris; Thomas Muffet (1553–1604), the entomologist whose daughter is forever frightened by a spider in a nursery rhyme; the physician and medical philosopher, Peter

Severinus (1540–1602); and Francis Bacon – developed their own versions of his chemical philosophy and proved influential in their own right [178; 51; 176]. But perhaps the greatest of these was Joan Baptista van Helmont (1579–1644), a Flemish nobleman, who merits his own label, Helmontianism [166]. Earlier historians of science have tended to seize upon quantitative aspects of some of his experiments (most famously that in which he weighed a pot, earth and a willow sapling before and after five years of watering, and concluded that the increased weight of the tree – 164 lbs – must have come solely from the water), while at the same time showing a whiggish exasperation with his all too obvious magico-religious outlook. The fact remains, however, that he was an extremely influential thinker, regarded in his own day as a leading representative of the new experimental approach to an understanding of the world. He was, for example, a major influence on that most representative of experimental philosophers, Robert Boyle [28].

It has often been suggested that the rise of the experimental method directly stimulated the formation of collaborative groupings of natural philosophers and practitioners of the various natural sciences, in more or less formal associations, such as the Accademia del Cimento (founded in 1657), the Royal Society of London (1660) or the Parisian Académie Royale des Sciences (1666) [163; 16; 152; 118; 96]. The assumption underlying these suggestions is that the experimental method demands collaborative effort, and certainly this claim was made explicit by Francis Bacon, whose ideal scientific institution, known as Salomon's House, described in his utopian *New Atlantis* (unfinished but published posthumously in 1627), was acknowledged as the inspiration behind both the Royal Society and the Académie des Sciences [116; 96]. Furthermore, the most eminent and successful of these associations, the Royal Society and the Académie des Sciences, can be seen to have grown out of less formal groupings of experimentally inclined thinkers, such as the Oxford and London groups which feature in the prehistory of the Royal Society [116; 228], or Montmor's Academy and the subsequent meetings under the patronage of Melchisédech Thévenot (1620–92) which feature prominently among the immediate causes of Colbert's decision to set up the Académie [16; 163].

33

Recent research suggests that this provides only part of the explanation for the sudden appearance of scientific societies during the Scientific Revolution. James E. McClellan has drawn attention to the fact that the formation of learned societies seems to have been an important feature of the Renaissance of learning in general and to go far beyond the confines of the natural sciences, embracing philology, literature, history and even theology. It seems that such societies developed as arenas for advanced, innovative work. In short, they were proto-Research Institutes, at a time when universities were merely teaching organizations [142].

Given this general background it is hardly surprising that each scientific society should have different origins. Clearly, the specific nature of their patronage is bound to shape the societies and the kind of work they did [212; 152; 141]. A number of the major differences between the Royal Society and the Académie, for example, can be attributed to the fact that the former was a self-supporting organization of interested members, while the latter was a carefully selected elite group, each of whom took a salary as a servant of the crown [118; 116; 96]. The importance of recruiting new fee-paying fellows into the Society meant that its projected image of Baconian collaborative science was necessarily somewhat looser than the more elitist Académie's view. The Royal Society needed to insist upon the validity of amateur contributions, the Académie was self-consciously a more professional body [96; 116]. Nevertheless, it remains clear that the scientific societies, the enthusiastic correspondence of their members, and their publications (such as the one-off *Saggi di Naturali Esperienze*, 1667, of the Accademia del Cimento, and the regular *Philosophical Transactions of the Royal Society*, from 1665) [163] did much to promote the new empirical method of practising science and of establishing natural philosophical truths.

Until very recently historians took too much at face value the dismissive critiques of some scientific reformers for the universities of their day. But the balance is now being redressed. Certainly, there was a lot of inertia in the university system, official curricula were slow to change, as were methods of teaching, and yet there is ample evidence to suggest that in some universities at least, curriculum notwithstanding, the latest ideas about the natural world, and about scientific method were being taught [86; 69]. Strict adherence to discipline boundaries between mathematics and natural philosophy within the universities clearly was a factor, as we've seen, in the lukewarm reception of Copernican theory and in Galileo's decision to seek private patronage as a natural philosopher, rather than remain a humble mathematics lecturer in a university [238; 13]. But the intellectual status of mathematics increased within universities just as it increased in the eyes of princely patrons. The importance of mathematics was extolled in the German universities where Melanchthon's pedagogical reforms held sway [238], the Jesuit *Ratio studiorum* also elevated mathematics [50; 109; 138], as did the proposed pedagogical reforms of the humanist Peter Ramus (1515–72) who was influential in the Netherlands [86]. In general, the increased recognition of the usefulness of mathematics led to improvements and increased opportunities in teaching it throughout Europe [191; 86; 69].

The experiential approach to an understanding of the physical world was, to some extent at least, always promoted in the Medical Faculties. The Italian universities, Montpellier in France, and even the highly traditional Paris Medical Faculty expected medical students to study practical aspects of medicine by a kind of apprenticeship to a local practitioner, while undertaking their more theoretical studies in the university. From the sixteenth century medical schools became the prime sites for a number of facilities essential for the promotion of observational and empirical science: anatomy theatres, botanical gardens, and in some cases chemical laboratories. Although the revolution in celestial and terrestrial mechanics might have proceeded largely outside the universities, the revolution in the life sciences took place almost entirely within a university context [191]. Medical Faculties became the prime sites for propagating the latest ideas in natural philosophy and for providing access to new, highly expensive instruments, such as microscopes, telescopes, and air-pumps. Cartesianism replaced Aristotelianism in a number of Dutch universities, while Paracelsian and other chemical philosophies were absorbed into the curricula in some German universities [86; 156].

Although it would be wrong to discount the role of the universities in the Scientific Revolution, it is important not to overstate the case. It should be borne in mind that, throughout this period, the function of the university was to teach. The sites for new research were the courtly Academies, the Royal Society, or the private house of a dedicated individual, whether a wealthy grandee like Tycho Brahe or Robert Boyle, or a more humble seeker after knowledge, like Andreas Libavius (1560–1616) or Antoni van Leeuwenhoek [86: 252; 72; 105; 197].

It seems reasonable to assume, then, that roughly contemporaneous developments in the mathematical sciences, natural history, physiology and anatomy, chemistry, and a concomitant development in instrumentation, all played their parts in the rise of empiricism at this time. It is clear also that increased awareness of the power of the experimental method led to new interactions between men of science, which stimulated further empirical investigation and, in turn, led to a formalization of association in scientific academies or societies. All of these developments were stimulated by, and in turn reinforced, changes in princely courts and universities. One general result was a radically altered belief in the authoritativeness of knowledge produced by experience. In these ways the new experimental method became a characteristic feature of the Scientific Revolution.

But this is to explain the rise of experimentalism as a routine accepted practice in the investigation of nature. Which is not quite the same as explaining the historical origin of what might be called *the* experimental method. What is usually meant by the 'experimental method' today is an artificial procedure performed in a laboratory to test a highly specific hypothesis within a credited theoretical framework. It will probably depend upon the use of special apparatus, in many cases specially designed and made for this particular experiment. It will also be designed in order to exclude, as much as possible, all other variables except the one which is being tested. It will be, at least in principle, endlessly repeatable, so that results can be checked time and again, or so that the effect can be demonstrated to new onlookers. It is precisely this 'experimental method' which allows scientists today to lay claim to their immense cognitive authority.

It is clear, however, that by no means all scientists can embody this method in their work (consider those working in the classificatory sciences of botany and zoology, for example). Furthermore, sociologists of science have repeatedly shown that scientists who might, in principle, live up to the demands of this method, in practice do not do so (even though they may retrospectively claim to have done so) Philosophers of science, moreover, have repeatedly been forced to acknowledge the impossibility of demarcating science from non-science in terms of a characteristic methodology. Furthermore, very little historical research is required to show that talk of a single, easily characterized experimental method is simply too glib. The experimental method of Harvey was not like that of Galileo, and neither's were like that advocated by Bacon, or that adopted by Robert Boyle. So, how is it that there is such a powerful conception of something called *the* experimental method which does such sterling rhetorical service in promoting the intellectual authority of science? Some recent work in the history of science has shown how historical research can play a major role in helping us to answer this question, while also providing us with a more precise understanding of the historical emergence of experimentalism.

In a detailed study of the attempts of Jesuits like Christoph Clavius (1537–1612), Orazio Grassi (*c.*1590–1654) and others to raise the status of mathematics, Peter Dear has shown that initial difficulties arose from the fact that mathematically produced knowledge claims were not self-evidently true [50]. The ideal of science in the Aristotelian tradition was based on the form of the logical syllogism, but the premises, the starting points upon which the reasoning was based, had to be uncontentious, evident truths to which all could freely assent. This was problematic for the mathematical sciences. In astronomy, for example, there were evident truths, such as the rising and setting of the sun, but the speeds of the planets, their retrograde movements and numerous other phenomena could only be established by skilled observations. Similarly, in optics, many phenomena could only be made apparent by experimental manipulations with special apparatus. To give the mathematical sciences the status of Aristotelian natural philosophy, these artificially produced observations had to be made to look evident to a wider public. Although there are a number of cases of experiments being performed in highly public places (such as dropping weights from the top of church towers), the most powerful means of making experiments 'public' was by developing new ways of describing them in published literature. The usual model of description was borrowed from geometry textbooks. The reader was instructed in how to set up the experimental scene and how to perform it, and then told what would ensue. For good measure, it became customary to claim that this experiment had been previously repeated a number of times, and often performed in front of various expert, named witnesses.

Dear has traced the influence of these developments upon the characteristic kind of experimental science performed on the Continent, which is very different from the experimental philosophy as it was performed in England [48; 50: Ch. 7]. When Blaise Pascal, for example, described an experiment it was presented in the form of a universal statement about how things happen. If you do this, this and this, then this will happen. Robert Boyle, leading light among English experimental philosophers, took exception to this. It looked to him like a report of what must happen, assuming that Pascal's theoretical assumptions were correct. Like Bacon, Boyle believed it was always possible to set up an experiment which seemed to confirm the experimenter's preconceptions. The experimental method which held sway in England, promoted by Boyle and a prominent group within the Royal Society, was intended, so they claimed, merely to establish the matters of fact. The English method, therefore, was held to be free from any bias introduced by theoretical preconceptions.

The rhetorical emphasis on 'matters of fact' in English natural philosophy has been brought out most forcefully in the work of Steven Shapin and Simon Schaffer [201]. In their important study of Boyle's efforts to establish the experimental philosophy as a means of determining truth and settling all dispute in natural philosophy, Shapin and Schaffer have shown how Boyle, and like-minded thinkers in the Royal Society, insisted that they were concerned only to establish matters of fact in their experiments, not to interpret their findings in accordance with any one of a number of alternative theories. Boyle's investigations with the newly invented air-pump were not intended, for example, to decide between the theories of those who believed in the possibility of void space, and those who did not, but merely to establish the springiness of the air. The rhetorical insistence upon the matter-of-factness of their experimental conclusions led the English experimenters to present their experiments as actual historical events. This led to the development of a new style of writing about experiments, to give the reader a sense of having been there. The purpose of this was to multiply the witnesses to the actual events, by making them 'virtual witnesses'. This was one way of getting around the problem of testimony: Why should these reports be trusted? The virtual witnesses were made to feel they knew so much about the experimental scene and procedure that they effectively witnessed it themselves. Otherwise, there was an appeal to the *number* of actual witnesses, typically at a meeting of the Society, far outnumbering the requirements of legal proceedings; or an appeal to the *nature* of the witnesses, typically gentlemen who spoke and acted freely and disinterestedly [201; 200; 197].

The attempt to replace the authority invested in the Aristotelian syllogistic approach to natural philosophy by authoritative experiments was not an abstract exercise in epistemology. The supposedly self-evident nature of the premises which gave traditional Aristotelian natural philosophy its authority, had to be replaced. Experiments, like mathematics, are not self-evidently true. To be convinced of their truth you either have to know what you are doing, or accept them on faith. Since it was as impossible for Boyle or Pascal to make everyone experimenters as it was to make them mathematicians, they concentrated on emphasizing the trustworthiness of their claims. But why the differences between them, between Continental experimentalism and the English kind?

Dear offers an explanation in terms of religious differences. Continental Catholics still believed in the occurrence of miracles, while English Protestants insisted that the age of miracles was passed and so no genuine miracles now took place. Dear sees, as a corollary to these beliefs, a Catholic belief in the fixed, law-like order of nature which can be violated by a single historical event (a miracle). A description of a single experiment in this setting is meaningless (it is merely another example of the order of nature), but a universal statement about, say, air-pressure actually reveals something about the order of nature. In Protestant England a belief in a fixed order of nature is not required to provide a benchmark for deciding what is miraculous, because miracles don't happen. Single experiments are meaningful, therefore, in contributing to a more precise understanding of the way things happen to be, whereas a universal statement is meaningless because it is based on what the majority of English Protestants take to be the false assumption that things could not be otherwise (vacua must exist, or the minute parts of bodies must be indivisible, or similar) [48; 50: Ch. 7; cf. 110; 112].

Shapin and Schaffer, by contrast, explain the emphases in the Royal Society's methodology in terms of the troubled history of seventeenth-century English society and the continuing need, after the restoration of the monarchy, to guarantee settlement and peace. Boyle and his colleagues believed that by concentrating upon the establishment of matters of fact they were providing the means to end dispute in natural philosophy. Everyone could agree about matters of fact even if they could not agree as to whether matter must be infinitely divisible or not. The united community of natural philosophers could then contribute to the establishment of order in society; legitimation of their method in and by natural philosophy meant that it could be used to lay down rules for the production of authentic knowledge, and the management of dispute in other areas, such as politics and religion. The English experimental method presented itself as a means for generating and maintaining consensus in a self-ordering community without any arbitrary authority [201: 341].

Some of the interpretations of Dear, Shapin and Schaffer are controversial, but their work goes a long way to helping us understand the precise nature of the 'experimental method' in seventeenth-century England, as contrasted with that more typical on the Continent. Moreover, they provide valuable materials for helping us to understand the power of the experimental method in the constitution of modern science. As Shapin and Schaffer point out, there is a tendency today to assume that the success of the experimental method requires no explanation because it seems to us to be so obviously superior to other ways of generating knowledge. Their historical analysis shows that in fact our present view of the validity and efficacy of experimentalism has its origins, like the experimental method itself, in various social, political and rhetorical strategies used in the early modern period for various local, historical purposes.

3 Magic and the Origins of Modern Science

Further important sources of the empiricism of the Scientific Revolution were to be found in the magical tradition, and these influences can be seen at work in a number of areas. They deserve separate consideration here, however, because they have generated considerable historiographical debate [222; 39]. A number of historians of science have refused to accept that something which they see as so irrational could have had any impact whatsoever upon the supremely rational pursuit of science. Their arguments seem to be based on mere prejudice, or on a failure to understand the richness and complexity of the magical tradition.

The Renaissance recovery of ancient texts which stimulated so many other areas of intellectual life clearly resulted in a renewed burgeoning of magical traditions. The flourishing of magic certainly owed a great deal to the rediscovery of ancient Neoplatonic writings, which included the writings attributed to Hermes Trismegistus (although it is now generally agreed that the claims of Frances Yates and her followers about the influence of the so-called Hermetic tradition have been greatly exaggerated [39]), but it is now clear that it also owed much to new trends within Renaissance Aristotelianism. The principles upon which the medieval theory of magical interactions was based derived, inevitably, from scholastic Aristotelianism. During the Renaissance a number of Aristotelian philosophers, notably Giovanni Pico della Mirandola (1463–94) and Pietro Pomponazzi (1462–1525), refined the more naturalistic aspects of the magical tradition (in which magical effects were brought about by exploiting the natural, but occult, properties of things), while Neoplatonizing philosophers, like Marsilio Ficino (1433–99) and

Tommaso Campanella (1568–1638), developed a more spiritual or demonic form of magic [38].

If we wish to understand the role of magic in the Scientific Revolution it is important to note the existence of so-called natural magic as, arguably, the dominant aspect of the magical tradition. Natural magic was based on the assumption that certain things had hidden, or occult, powers to affect other things and so accomplish inexplicable phenomena. Success as a natural magician depended upon a profound knowledge of bodies, and how they act upon one another, in order to bring about the desired outcome [38; 111; 230]. Repeatedly we see Renaissance natural magicians insisting that their form of magic depended upon nothing more than knowledge of nature, so much so that one recent historian has suggested that we should designate this kind of thinking as 'Renaissance naturalism' to distinguish it from what he thinks of as real magic [114].

In a very real sense, however, the separation of the naturalistic elements from other aspects of magic was just what was accomplished during the Scientific Revolution. The history of magic since the eighteenth century has been the history of what was left to that tradition *after* major elements of natural magic had been absorbed into natural philosophy. Moreover, for us magic deals with the supernatural, but for early modern thinkers magic relied for its effects upon the manipulation of natural objects and processes. For them only God could bring about supernatural events. Even the demonologist, in summoning a demon – perhaps the Devil himself – to do his bidding, only expected the demon to be able to perform like an extremely knowledgeable natural magician, using the hidden *natural* powers of objects to bring about desired events [25; 38; 114]. The reason why natural magic has disappeared from our conception of magic is precisely because the most fundamental aspects of the tradition have now been absorbed into the scientific world-view. Or, to put it another way, the scientific world-view developed, at least in part, out of a wedding of natural philosophy with the pragmatic and empirical tradition of natural magic.

The pragmatism of magic is obvious. The aim of the magus is always to bring about some desired outcome, either for his own benefit or for that of a patron or client [39; 214]. The empiricism of magic may seem more surprising to us. Yet this

43

is built into the logic of natural magic. How else could the magus learn about the occult powers of one body to affect another? The less diligent might rely on semiology: reading the signs, or *signatures*, which God has set down to enable us to read the book of nature (a favourite historians' example is the walnut, the structure of which resembles the brain inside the skull, a clear sign from God that it can be used to cure diseases of the brain). The more diligent will check things out for themselves (although, in practice, natural magicians tended to rely heavily upon traditional claims in the magical literature; if something is mentioned in more than one or two books, it must be authoritative! [38]).

The relevance of magic to the reformation of ideas about the correct way to understand the natural world can be seen in the fact that, surprising though it may seem, technology was inextricably linked with magic throughout the Middle Ages and the Renaissance [61; 62]. This does not mean that the uninitiated believed machines were worked by inner demons (remember, magic was believed to work by *natural* means). The elaboration of mechanical contrivances to produce marvellous effects was simply regarded as the exploitation of the occult, but natural, powers of things and therefore the province of the magician [61; 216]. Moreover, because of the close links between mechanics and mathematics, this kind of exploitation of machinery was often called 'mathematical magic'. So magic also became associated with the mathematical approach to understanding the physical world [62; 111; 29; 68].

Having said that, however, it is necessary to be extremely careful in assessing just how mathematics was said to reveal the secrets of nature. It is clear, for example, that the significance of numbers meant different things to different thinkers. Johannes Kepler, consummate mathematical astronomer that he was, can also be seen to have been deeply affected by the magical tradition of numerology. It is well known that a major stimulus to his work in cosmology was his attempt to answer the question why there were only six planets. This was not a scientific question, it sought to understand what was so significant about the number six, that God should have used it, and no other number, to be the number of planets. But there is a world of difference between Kepler's numerological question

44

and how he pursued the answer and, say, Robert Fludd's kind of numerology. Fludd (1574–1637) was a prolific writer of magical works of the most mystical kind. Kepler objected that the numerical ratios in the heavens which Fludd discussed in his *Utriusque Cosmi Historia* (1617–21, *History of Both Cosmoses* – Fludd means the macrocosm and the microcosm), were mere symbols, dreamed up by Fludd to serve his poetical and rhetorical purposes. Kepler insisted that the numbers and numerical ratios with which he himself was concerned were real features of the physical world. In other words, the numbers Fludd used were imposed upon the heavens by his own fancy, but Kepler used only numbers which could be seen to be built into the actual system [39; 71; 53; 99]. Kepler's distancing of himself from Fludd in this way should not be seen as a rejection of magical traditions, however, but as a reaffirmation of sound natural magic. When Francis Bacon wrote that 'There is a great difference between the Idols of the human mind and the Ideas of the divine', he might have spoken for Kepler, and it is clear from what he went on to say, that he nonetheless believed that God's Creation bore signs of its significance and usefulness for mankind: 'That is to say, between certain empty dogmas, and the true signatures and marks set upon the works of creation as they are found in nature' [quoted in 8: 323].

Accordingly, it is important to note that Kepler's determination of celestial harmonies, using Tycho Brahe's extremely accurate observations, enabled him to prove to his own satisfaction that there were only six planets because God created the universe in accordance with a 'geometrical archetype'. God separated the planets from one another, Kepler became convinced, by nesting them alternately with each of the five so-called Platonic solids. The point about these solids is that the principles of Euclidean geometry establish that they are the only three-dimensional bodies of their kind. No other closed solid can be made with all faces the same. Thus, when God inscribed a cube inside the sphere of Saturn, touching the sphere at its eight corners, and then used that cube to demarcate the sphere of Jupiter, so that that sphere touched each face of the cube, and proceeded similarly with the tetrahedron between Jupiter and Mars, and so on, the creation of planets had to stop with six, since there were no remaining solids which God could use

God, shapes, magic ... numerology.

45

[71; 134]. Kepler's use of mathematics was very different from Fludd's, but it can hardly be said that it was not steeped in the Neoplatonic magical tradition.

Natural magic was, as William Eamon has pointed out, 'courtly science par excellence' [62: 225], and it flourished in the courts of Europe, particularly in the earlier part of the period [64; 65; 154; 155]. The earliest of the courtly academies which concerned themselves with natural knowledge can be seen to be instituted for the advancement of natural magic. For example, when Federico Cesi (1585–1630) founded the Accademia dei Lincei (Academy of the Lynxes) he was inspired by Giambattista Della Porta's (1540–1615) discussion, in the preface of his compendious *Magia Naturalis* (1589), of the need to observe nature with lynx-like eyes in order to put natural things to use [62: 229–33]. In the universities, astrology always figured quite largely, especially in the medical faculties [214], but in the sixteenth and seventeenth centuries other aspects of the magical tradition made their presence felt, in particular other aspects of mathematical magic and alchemically inspired medical theories [68; 154]. The magical belief in signatures and the correspondences between different rungs in the ladder of creation have been seen as a major stimulus to the careful observation and recording of minerals, plants and animals [175: Ch. 8; 64: 245–6; 8]. Not even the new natural philosophical instruments escape the taint of magical antecedents. Deceptive tricks with lenses and mirrors had always been among the more dazzling of the natural magician's arts, and the telescope and microscope were treated with extreme caution by most natural philosophers when they were first introduced into the study of nature [219; 220; 104].

It seems undeniable that magical traditions played an important part in the major shift from scholastic natural philosophy to the new, more practically useful, more empirical, natural philosophy of the Scientific Revolution. The precise details of how some aspects of the magical tradition were taken up and others vigorously rejected remain far from clear, however. Presumably, part of the story was dictated by increasing awareness among patrons and practitioners as to what methods were most efficacious, what underlying assumptions pointed the way to the most fruitful conclusions, and so forth. Since magic had always had a bad public image, deriving chiefly from the prevalence of fraud among self-proclaimed magicians and from the unceasing attacks of the Church, it made sense for reforming natural philosophers to add their own voices to the denunciation of magic, while they extracted what they recognized to be useful out of the tradition. Something of this attitude can be seen at work, for example, when Seth Ward (1617–89) disputed with John Webster (1610–82) in 1654 about the lack of magic in the university curriculum. Ward dismissed magic as a 'cheat and imposture' which deludes 'with the pretence of specificall vertues, and occult celestiall Signatures', but immediately insisted that 'The discoveries of the Symphonies of nature, and the rules of applying agent and material causes to produce effects, is the true naturall Magick, and the generall humane ends of all Phylosophicall enquiries' [52: 228–9]. Perhaps the best example of this duplicity towards magic is Francis Bacon who, while drawing much of the inspiration for his new method from the magical tradition and developing what has been described as a 'semi-Paracelsian cosmology', managed to distance himself from magic by vilifying it as much as anyone [176; 180]. When Bacon set down the aim of the idealized scientific Academy, which was the focal point of his utopian *New Atlantis*, he used the very language of the natural magician: 'The End of our Foundation is the knowledge of Causes, and secret motions of things; and the enlarging of the bounds of Human Empire, to the effecting of all things possible', and he drew freely upon magical sources for much of the material in his encyclopaedic *Sylva Sylvarum* (1627). It is clear from his criticisms that he saw many faults in the magical approach, but it cannot be denied that his own work was greatly affected by the magical tradition [180].

The lesser traditions of Paracelsianism [51; 54; 229], Helmontianism [166] and derivative chemical philosophies suffered similar fates. A number of Paracelsian ideas became absorbed into mainstream medicine, chemical remedies appeared in the official pharmacopoeias, but Paracelsus himself and his followers were frequently vilified. Because of the radicalism of his break with traditional Galenic medicine, Paracelsus came to be seen as the Luther of medicine, but this meant that Paracelsianism could not be adopted lightly. Galenism was entrenched within the medical schools and the licensing authorities, the colleges of physicians, demanded orthodox Galenism from licensed practitioners. Like Aristotelianism, Galenism was seen as another aspect of traditional authority, to embrace Paracelsianism could be seen, therefore, as a sign of subversion. Certainly Paracelsianism flourished in societies rent by religious and political factionalism. In late sixteenth-century France Paracelsianism was promoted by the Protestant Huguenots, in the early seventeenth-century it flourished in Protestant German states, particularly in Bohemia before it was crushed by the Holy Roman Emperor, Ferdinand II, in his attempts to re-establish Catholicism [178; 54]. Subsequently, Paracelsianism thrived in England under Parliamentarian rule when the College of Physicians was seen as a 'Palace Royal of Galenical Physick', and Galen as a tyrant in medicine to be deposed like Charles I [178; 228; 35; 149; 174]. Inevitably, in view of its radical affiliations, Paracelsianism went into decline in England after the Restoration, although it managed to leave its mark on the practice of medicine. Only a meticulous study of changes in pharmacopoeias and in medical practice can reveal which elements of Paracelsianism and other chemical philosophies began to appear on the 'scientific' side of the boundaries newly erected in the Scientific Revolution, while others remained in outer darkness. Moreover, only a study which takes into account the social, religious and political background can explain why the boundaries were erected where they were. In the meantime, however, it seems entirely justified to note that the magical tradition played an important role in the establishment of empirical and pragmatic attitudes to natural knowledge.

But this is not the whole story. Magical influence was not confined to general, methodological matters. There are a number of cases in which substantive conceptual innovations can be shown to owe a great deal to magical ways of thinking. Nor are these merely marginal innovations. Magical conceptions can be seen to play an important part in the thought of a number of leading thinkers. Leaving aside the chemical philosophers, for whom no case has to be made, the list would have to include William Gilbert, Johannes Kepler, Robert Boyle and Isaac Newton.

Consider William Gilbert, whose experimental investigations of magnetism have been seen as foundational. Edgar Zilsel and J. A. Bennett [245; 9] point to the socio-economic importance of the magnetic compass and the pragmatically inspired investigations of navigators, mariners and the like, as a major stimulus towards Gilbert's investigations and a major source of his method. But even a superficial reading of Gilbert's *De Magnete* (1600) is sufficient to reveal his animistic and magical approach to the natural world. For Gilbert the earth was an animate body, capable of moving itself in the same way that a magnet, always regarded as the supreme example of a magical object, could move itself (he was writing to a Copernican agenda: seeking to account for the motion of the earth). Many of his experiments were concerned to establish the spontaneous movements of magnets with a view to showing that they must, therefore, be possessed of souls. He even went so far as to suggest (and Platonic influence is clear here) that the magnetic soul was superior to the human soul because it was not deceived by the senses, as the human soul all too often was. Subsequent experiments were conceived to prove that the earth itself was a giant magnet, thus making it an easy step to the conclusion that the earth had a soul (and was therefore capable of self-movement).

Johannes Kepler adapted Gilbert's ideas in his physicalist *New Astronomy* (1609), explaining the motions of the planets around the sun by recourse to something like magnetic force, but magical traditions can be seen to be much more prevalent in his thinking elsewhere in his writings. We have already noted Kepler's attempt to explain why there were only six planets, but he was also concerned to know why the planets were placed where they were. This was puzzling for Kepler in so far as they were placed with no obvious pattern, instead of being, say, evenly spaced. Incredible though it may seem, Kepler's geometrical archetype, in which he nested the five Platonic solids between each of the planetary spheres, actually provided an impressively accurate answer to that question. The Platonic solids not only determined that there could only be six planets (because not even God could make another closed solid with all faces equal), but they predicted the spacings that the Copernican astronomer could establish by geometrical calculations – or very nearly so [71]. Nor did he abandon the geometrical archetype, which relied upon the notion of planetary *spheres*, after his discovery of elliptical orbits. In order to explain why God should use ellipses instead of circles (or spheres), Kepler drew upon the Pythagorean and Neoplatonic tradition of celestial harmonies. Planets moving in circles with unchanging speed could only generate monotones, he reasoned, but a planet moving with regularly varying speed on an ellipse would generate a range of notes. In trying to work out the precise notes made by each of the planets, Kepler used Tycho's accurate observations to determine, among other things, the speeds of the planets when closest and farthest from the sun. Kepler's so-called third law of planetary motion, which gave a precise relationship between the time taken for a planet to complete a full circuit of the heavens and its *average* distance from the sun, enabled him to return to his geometrical archetype. The average distances of the planets from the sun provided a set of circles (or spheres) which could once again be shown to fit in place between the Platonic solids. Astonishingly, even with Tycho's planetary observations (not significantly different from modern ones), the accuracy of the fit was remarkable. Small wonder that Kepler believed he had discovered the blueprint God used when creating the cosmos [71: Ch. 5; 14; 210].

The same tradition of Pythagorean or Neoplatonic cosmic harmonies can be seen at work in the writings of Isaac Newton. In some manuscript drafts which he composed for inclusion in an abandoned second edition of the *Principia Mathematica*, we see Newton indulging in magico-religious speculations about an esoteric knowledge of universal gravitation among the ancient followers of Pythagoras (who, by the way, was regarded as a leading figure in the genealogy of magic; the badge of his brotherhood, supposedly chosen because of the mathematical nicety that all its lines divided one another in the so-called golden ratio, was the pentacle, which became a well known symbol of the magician). The Pythagorean doctrine of the harmony of the spheres, symbolized by means of Apollo, the sun-god, with a lyre of seven strings, indicated their belief, according to Newton, that the Sun attracts the planets in accordance with the inverse square law [144: 116]. The complete extract makes it look as though Newton was convinced of the numerological significance of the number seven, and nor is this the only place in his writings where the number seven plays an important role. His discussion of the colours of the spectrum, both in his early paper to the Royal Society (1675) and in the *Opticks* (1704), draws a precise analogy between the colours and the seven notes in the octave. Newton even claimed that repeated experiments, in which the positions of the projected colours were marked on sheets of paper, showed the agreement between these marked distances and the positions needed to bridge a monochord of corresponding length to produce the notes of the diatonic scale. In his unpublished lectures on optics, however, we learn that Newton first measured the distances between five colours and subsequently added orange and indigo [92]. It would seem that when British schoolchildren learn the *seven* colours of the rainbow they are paying unwitting homage, not to Newton's experimental method, but to his belief in cosmic harmonies.

Newton was also an alchemist. For a long time dismissed by historians as irrelevant to an understanding of his scientific endeavours, his alchemy has more recently been seen as an important element in his thinking about the nature of matter. Betty Jo Dobbs and R. S. Westfall have insisted that Newton's willingness to let his system of physics depend upon occult forces of attraction and repulsion operating between the particles of matter stemmed from his familiarity with alchemical modes of thought [56; 237]. It has been pointed out, however, that Newton first relied upon such forces, capable of operating across empty space, after Robert Hooke suggested to him, in 1679, that Kepler's laws of planetary motion could be explained in terms of a planet's linear motion bent into an ellipse by an attractive force towards the sun, varying inversely as the square of the distance between sun and planet (see Chapter 2.i) [31]. Since Newton had been an alchemical adept for a number of years before this, it seems impossible to dismiss the significance of Hooke's hint, especially in view of a recent claim that in order to understand Hooke's own thinking on this matter, we need look no further than the contemporary scene in mainstream English natural philosophy [110].

But if alchemy cannot be said to have been the prime inspiration behind Newton's belief in occult forces, it can be seen to have profoundly affected other aspects of his scientific thought. From his early 'Hypothesis of Light', sent to the Royal Society in 1675, to the speculative 'Queries' which he added to successive editions of his *Opticks* (3rd edition 1717), Newton clearly drew upon alchemical conceptions about the activity of light and its ability to interact with and bestow activity upon matter. If Newton had the idea for an occult force of gravity from elsewhere, his speculations about active principles in matter, one of which might be the cause of gravity, seem to have come straight out of the alchemical tradition [57]. There is no getting away from the fact, either, that Newton's gravity was an occult force, capable of acting at a distance across vast expanses of space. Even if this concept did not emerge from Newton's alchemical researches, it still testifies to the importance of magical traditions in his thinking [56; 237].

Newton's older contemporary, Robert Boyle, certainly the most respected natural philosopher in England in his day [201; 200; 120], was also a practising alchemist and the theory underlying his alchemy can be seen to have shaped some of the details of the natural philosophy for which he is revered. Boyle's corpuscular philosophy, which has been seen as deriving ultimately from the philosophy of Descartes, or from the reviver of ancient atomism, Pierre Gassendi (1592–1655), is in many respects much closer to a particular alchemical tradition, stemming primarily from the *Summa Perfectionis*, falsely attributed to the Arabic alchemist known as Geber [159]. Other aspects of his natural philosophy can be seen to stem from more common features of the alchemical tradition [171; 27], and yet others from the chemical philosophy of Joan Baptista van Helmont [28].

The Renaissance revival of the natural magic tradition made another crucially important contribution to the Scientific Revolution. One of the major premises of natural magic was that some (if not all) bodies have occult powers capable of acting upon some or all other bodies. Typical occult qualities, acknowledged by all, were the different influences of the planets, magnetism, and the ability of certain minerals, plants, and even animals to cure various diseases. These occult powers were so-called because they were insensible; we cannot perceive the magnetic power by means of our senses, we only know of its existence by its effects; we cannot understand by inspection how rhubarb purges the bowels, but we need be in no doubt of its efficacy. In traditional scholastic Aristotelianism such occult qualities were something of an embarrassment; it was difficult to accommodate insensible causes in a natural philosophy based on explanation in terms of evident causes. The scholastic philosopher felt satisfied if he could explain changes in terms of the *manifest* qualities of heat, cold, wetness and dryness; recourse to occult qualities seemed to be an admission of intellectual defeat [121; 153].

In the medical tradition, however, where it was frequently recognized that the powers of drugs could not be explained in terms of their manifest properties, occult qualities were much more routinely invoked [153; 38]. With the rapid expansion of the lore of flora and fauna which was a feature of the burgeoning of observational natural history, and with the development of chemical remedies and other advances in chemical knowledge, a whole new area of occult qualities was opened up. Aristotelian natural magicians, like Giovanni Pico and Pomponazzi, and later reforming Aristotelians, like Girolamo Fracastoro (c.1478–1553) and Daniel Sennert (1572–1637), sought ways to accommodate such occult qualities into Aristotelian natural philosophy. There were two main approaches. Firstly, to suggest some means of natural causation which, although insensible, was not unintelligible (examples included emanating subtle spirits, or the actions of invisibly small particulate effluvia). Secondly, to emphasize the reality of these occult qualities by pointing to the empirically undeniable reality of their effects. Here was another major stimulus to the empirical investigation of nature [121; 153; 38].

These ideas were to come to fruition in the new natural philosophy of the Scientific Revolution. When Francis Bacon famously rejected the deductive logic of the syllogism in favour of inductive logic, he was elevating the logic which was implicit, if not explicit, in the natural magic tradition [111]. The scholastic distinction between manifest and occult qualities was effectively dismissed by Bacon's method of gathering empirical facts and setting them down in 'Tables of Instances' (Bacon believed that only a pre-theoretical gathering of bare facts could guarantee that the explanation of a natural phenomenon would not be prejudged, or prejudiced) [145]. Incidents involving heat (for scholastics, a manifest quality), and incidents

Compatibilizing "occult" forces w/ Aristotelian physics.

involving magnetism (an exemplary occult quality), were to be dealt with in the same way; the result was that heat no longer seemed manifest (being deemed by Bacon an 'expansive, restrained' motion of the particles of bodies), and magnetism seemed no more unintelligible than heat [121; 153].

Although Bacon never managed to fully articulate his new inductive method before his death, he did succeed in convincing some of the subsequent generation of natural philosophers (particularly among his fellow countrymen) that the experimental method could be used to sanction the use of occult qualities in scientific explanations [110; 112]. The so-called 'experimental philosophy', as it was developed in England, allowed the use of unexplained physical phenomena provided that their effects could be made manifest by experimental means. Boyle and Hooke frequently explained pneumatic phenomena in terms of the 'spring' of the air. They eschewed hypotheses about the *cause* of the air's self-expansive endeavours, being content to insist upon its reality, as made plain by effects in the air-pump [110; 201]. This same Baconian tradition can be seen at work in Isaac Newton's confident rebuttal of the charge (made by Leibniz) that his principle of gravitation was a 'scholastic occult quality'. For Newton, although the cause of gravity remained occult, gravity itself could be said to be a manifest quality, because of our daily experience of it and because of his precise mathematical analysis of its operations [110; 112].

Of the two responses of Renaissance Aristotelian natural magicians to the contemporary proliferation of occult qualities which seemed to them to be a consequence of developments in natural history, chemistry, and other arts, English natural philosophers certainly focused more upon the way of experimental ratification. On the Continent, however, the emphasis was much more upon the attempt to find intelligible causal explanations for the insensible *modus operandi* of occult qualities. In order to understand this difference of emphasis we must look to differences in the social and political background. While various hypotheses about underlying intelligible causes simply led to dispute and conflict in natural philosophy, Bacon's method of gathering facts to establish empirically the reality of occult qualities could be made to serve the irenic aims of Boyle and others, seeking to present a natural philosophy which could

54

smooth over disagreements and command general assent. This aspect of the natural magic tradition can be seen, therefore, to fit into the reforming philosophical, religious and, ultimately, political ambitions of Boyle and like-minded English contemporaries which have been described by Steven Shapin, Simon Schaffer and others [201; 203; 243; 46]. Furthermore, something of the same kind of irenicist motivation in English natural philosophy can be traced back to the beginnings of the century [112].

Once again, therefore, we can see that the experimental method, and indeed the particular English version of it, with its emphasis on Baconian fact-gathering and its self-professed rejection of speculative theorizing, derived in large measure from the natural magic tradition. By the same token, the natural magicians' alternative way of accommodating occult qualities in natural philosophy, by putative insensible but physical means, can be seen to have been influential in the development of the new systems of mechanical philosophy [121; 153] which are another salient feature of the Scientific Revolution, and which we consider in the next chapter.

The idea of things having invisible qualities was a new one.

4 The Mechanical Philosophy

Natural philosophy and the mathematical disciplines underwent considerable reforms during the Renaissance but before the dominant scholastic-Aristotelianism could be replaced something more was required. Scholastic natural philosophy was a complete system, seemingly capable of dealing with most questions about the physical world. The Aristotelianism which formed the core of the system was dovetailed pretty neatly with Ptolemaic astronomy and with Galenic medicine. Furthermore, it was based upon a coherent and powerful metaphysics, and, thanks to the work of Thomas Aquinas and other Church leaders since the thirteenth century, it was seen as a 'handmaiden' to the 'Queen of the Sciences', theology. The essential unity of approach to the nature of the physical world, from the macrocosm to the microcosm, was seen as unshakeable testimony to the truth of the system. During the Renaissance that unity began to break up, but the general tendency among intellectuals was to patch up the old system and to stick with it. To be a natural philosopher, after all, was to be in possession of a key to answering all questions about the physical world. The result was, however, a proliferation of Aristotelianisms: a whole series of often ingenious refinements, re-workings and re-interpretations of traditional scholasticism to accommodate the latest findings and the latest fashions of thought [192].

While this state of affairs clearly satisfied many, there were others who wanted more. For them, what was required was a new system of philosophy, capable of replacing the Aristotelian system, root and branch. A number of rival attempts to produce such a system were developed, but for contemporaries they were generally regarded as versions of the 'mechanical philosophy'. By the end of the century the mechanical philosophy had effectively replaced scholastic-Aristotelianism as the new key to understanding all aspects of the physical world, from the propagation of light to the generation of animals, from pneumatics to respiration, from chemistry to astronomy. The mechanical philosophy marks a definite break with the past and sets the seal upon the Scientific Revolution.

In its strictest forms the mechanical philosophy was primarily characterized in terms of a restricted range of explanatory principles. All phenomena were to be explained in terms of concepts employed in the mathematical discipline of mechanics: shape, size, quantity, and motion. The logic of this kind of explanation tended to lead to a restricted theory of causation, conceived only in terms of contact action. The mechanical philosophy saw the workings of the natural world by analogy with machinery; change was brought about by (and could be explained in terms of) the intermeshings of bodies, like cogwheels in a clock, or by impact and the transference of motion from one body to another. Explanations in terms of animate principles, and teleological accounts (in which the behaviour of something was explained by reference to its supposed purpose: why does an acorn grow? because its purpose is to become an oak to supply mankind with wood) were rejected. A distinction was made between what were considered to be the real properties of bodies (size and shape, motion or rest) and merely secondary qualities, caused by the former, such as colour, taste, odour, hotness or coldness, and the like [55; 235]. It is significant that the manifest qualities of Aristotelianism are reduced to being secondary qualities, brought about by the motions of the invisibly small particles which are held to make up large bodies. Similarly, occult qualities are explained by recourse to mechanical principles. The Aristotelian distinction between manifest and occult qualities is no longer significant in the mechanical philosophy since all explanations resort, ultimately, to the motions and interactions of insensible particles [121; 153].

This brings us to the final major characteristic of the mechanical philosophy: it was based on the assumption that bodies were made up of invisibly small atoms or corpuscles. It is hardly surprising that one of the major sources of inspiration behind the formation of the new systems of mechanical philosophy was the revival of the ancient atomist philosophies of Democritus, and more especially, of Epicurus. Indeed, one of the major systems of mechanical philosophy, that of Pierre Gassendi, was based upon his attempts to reconstruct the natural philosophy of Epicurus [130]. But not all mechanists believed in the existence of necessarily indivisible atoms. It was possible to be a mechanist and subscribe to the belief that matter was infinitely divisible, while insisting that in practice there were basic minimal particles involved in all physical change. Robert Boyle, for example, referred to himself as a corpuscularist, never as an atomist [201].

Another major stimulus to particulate theories of matter developed within the Aristotelian tradition itself. Initiating principally in the influential Arabic commentaries upon Aristotle written in the twelfth century by Averroes (*c.*1150), and stimulated by an increasing awareness of chemical change in the Renaissance, the supposition that substances were composed of *minima naturalia* played an increasing role in scholastic speculations about the nature of matter. Eclectic Aristotelians like Daniel Sennert, Girolamo Fracastoro, and David van Goorle (*fl.* 1610) melded the tradition of *minima naturalia* with atomism in their attempts to reform medical and chemical theory [221; 147; 63]. One important aspect of these reforms was the establishment of a concept of particles which although held to be indivisible had a finite size. Earlier atomist theories were confused by what might be called mathematical atomism, in which atoms were held to be indivisible because they were nothing more than geometrical points, without any extension. But such non-extended atoms could not easily be conceived as taking part in physical explanations of extended entities; minimal but extended particles could be. Gassendi was able to draw support from the *minima* tradition in his own attempts to clarify the principles of Epicurean atomism [130; 221].

The major rival to Gassendi's Epicurean system in the first half of the seventeenth century was the new system of René Descartes. The Cartesian system was, arguably, the most influential version of the mechanical philosophy, and in many ways was the most impressive [87; 89; 207]. Based upon a unification of mathematics with physics, legitimated by a new metaphysics, Descartes' philosophy defined matter solely in terms of extension. This enabled him, in principle at least, to claim that physics could be based upon geometrical analysis of extended bodies in motion [88; 107]. In practice, however, Descartes' exposition of the system of the world is hardly ever grounded upon actual mathematical analysis. His account of celestial motions, for example, relates the density of planets to their distances from the sun, but there is no attempt to calculate this relationship precisely. Descartes' confidence about the mathematical certainty of his system is based upon the axiomatic structure of the system, its supposedly indubitable foundations and the careful deduction of phenomena from those foundations [107].

Descartes' first account of his system, *Le Monde*, was completed in 1633 but he withheld it from publication when he learned of Galileo's condemnation for advocating Copernican doctrines. The mature version of his mechanical philosophy was published in 1644 as the *Principia Philosophiae* (still Copernican but with some nifty casuistry about all motions being relative, which enabled him to define the earth as stationary). The identification of matter with extension, which forms the starting point of the system, entailed the denial of vacuum, and provided the foundation for the claim that all interaction was by contact action. Because the world is completely full, motion can only occur by a displacement which is likely to have world-wide ramifications. In order to avoid this absurdity, it is supposed that displacements typically occur in a fairly localized cyclical pattern. A moving body can be seen, therefore, to form a circulation of matter, a vortex, or a series of such vortexes, in the surrounding densely packed particles as it proceeds [86; 207; 235].

Drawing upon mechanical accounts of centrifugal motion (as seen familiarly in a sling-shot), Descartes was able to account for celestial motions and for gravity by means of this vortex theory. If we assume a vast vortex of matter, the smallest particles will accumulate at the centre because they have a lesser centrifugal tendency. The crowding and jostling of these particles at the centre will generate friction, causing the light and heat we see in the sun and the fixed stars (each of which is the centre of a vortex). Larger particles conglomerate together to form the planets which are carried around in the vortex in fixed orbits in accordance with their densities. If a planet approaches nearer to the centre of the vortex, for example, it will encounter smaller, more rapidly moving particles (bigger particles having a greater centrifugal tendency will be further from the centre), accordingly its own motion will be increased and it will acquire a greater centrifugal tendency which will carry it outwards again. The system is self-regulating. But, why do larger particles conglomerate to form planets? This is not properly explained, but once a planet is formed, it becomes the centre of its own vortex. The motion of the particles surrounding the planet provides a centrifugal tendency with respect to the planet which is called upon to explain gravity [2; 88; 89; 207; 233].

Another important premise of Descartes' system is the claim that the amount of motion in the world always remains constant. Descartes tries to establish the precise ways in which motions are transferred from one body to another in seven rules of impact, which are supposed to follow from his three laws of motion. On the face of it, there seem to be a number of inconsistencies in these rules. Descartes denies, for example, that a moving body, no matter what its speed, could set a *larger* stationary body in motion. The clear implication is that the larger body has some power to resist motion, but this is incompatible with another mainstay of Cartesian philosophy, that matter is completely inert. Matter cannot be completely passive and have the power to resist motion [78; 89; 207; 233]. In fact, Descartes himself seems to have believed that this rule was justified by his principle of the immutability of God and his first law of nature, which held that bodies will remain in the same condition of motion *or rest* if they can, but it evidently caused considerable difficulties for Descartes' contemporaries as well as his modern commentators [106].

Indeed, Descartes' belief in the constancy of the amount of motion in the world was problematic enough without engaging with the rules of impact. The implication was that there could be no new motion in the world. When motion started somewhere in the world, somewhere else in the world the corresponding amount of motion had to be absorbed. Consider the case of putting a match to gunpowder. Certainly the motion of the match and its flame could not be said to have provided the motion of the cannonball, so what did? and how was that motion transferred to the ball by impact action?

Extraordinarily speculative and unconvincing though these sample explanations may seem to us, Descartes was convinced of their truth, as he tells us at the very end of the *Principles*, because they were deduced in a continuous series (like the theorems of Euclid) from 'the simplest principles of human knowledge'. Furthermore, many of his contemporaries were convinced, for although his followers might criticize details of the system, they were nonetheless convinced that Descartes had found the most reliable and fruitful way to understand the physical world.

Descartes' writ ran large on the Continent, particularly in France and in the Netherlands [129], but it did not have the same success in England. The experimental philosophy as it was developed in England precluded an easy acceptance of any deductive system and Descartes' system was seen to be just as divisive in natural philosophy as the extremely materialist system of Thomas Hobbes [201]. Although Descartes did allow for the use of experiments in the development of his system they played a distinctly secondary role in support of the chains of reasoning [107]. Consequently, Cartesian experiment tended to look like a report of what must happen, on the assumption that Descartes' reasonings were correct. As we've seen (Chapter 2. ii), this kind of presentation of experiment was rejected by Boyle and other prominent experimental philosophers in England [48; 201; 200].

This does not mean that the mechanical philosophy did not thrive in England. All the mainstream natural philosophers after the Restoration can be seen to have been mechanical philosophers, but there were marked differences between their kind of mechanicism and what might be considered to be the stricter versions of Descartes and Hobbes. One way of characterizing the difference can be drawn from their different responses to the natural magic tradition. We saw in Chapter 3 how natural magicians dealt with occult qualities in one of two ways: either by seeking to explain occult effects in terms of intelligible, material principles, like invisible particles; or by taking an empirically sanctioned phenomenalistic approach, so that the reality of an occult quality could be affirmed by its observed effects. We saw also, that the second alternative seemed to suit the interests of the English natural philosophers who had other reasons for wishing to place emphasis on experimentally established matters of fact. The result was a version of the mechanical philosophy in which matter was not consistently considered to be entirely passive and inert, as it was declared to be in the systems of Descartes and Hobbes. It was held to be possible that particles of matter could be endowed with active principles which might account for occult phenomena like magnetism and gravity, and various chemical properties (including, for example, the explosive property of gunpowder), but which could still be dealt with in natural philosophy by means of experimental demonstration and manipulation [110; 28; 27; 112]. It is surely significant also that Gassendi was more influential in England than Descartes [110].

According to Gassendi atoms were endowed at the Creation with an internal principle of motion, a 'natural impulse', 'internal faculty' or 'force' which always maintained their motions or, in some unexplained way, maintained their power to move (he talks of the motive force in atoms being 'held back', when the body which they constitute slows down, and 'liberated', when a body begins to move) [20: 76–9, 119–21; 164: 191–3]. Gassendi's philosophy also seemed to offer a more plausible way than Descartes' of dealing with vital and chemical phenomena. Some atoms, at least, were held to possess an 'internal force' or 'seminal power' which enabled them to create the seeds of plants, or in the case of 'lapidific' and 'metallic' internal powers, stones and metals [20: 129–33]. Gassendi's natural philosophy was paraphrased in English by Walter Charleton (1620–1707) in his *Physiologia Epicuro-Gassendo-Charltoniana* of 1654, and became immensely influential on leading English thinkers like Boyle, Newton and the philosopher, John Locke [110]. Gassendi's philosophically rigorous restatement of Epicurean principles not only demonstrated in detail how atomism could be used to explain physical phenomena, but also made atomism theologically and morally respectable [164; 130].

The culmination of the English tradition of mechanical philosophy can be seen in Isaac Newton's *Principia Mathematica*

(1687), and in the Queries appended in increasing numbers to successive editions of his *Opticks* (1704, 1706, 1717). When Newton wrote in the 'General Scholium', added to the second edition of the *Principia* (1713), that induction from phenomena made it plain that 'gravity does really exist and act according to the laws which we have explained', even though the *cause* of gravity remained occult, he very much offended the Cartesian sensibilities of thinkers like Huygens and Leibniz.

For Huygens, the notion of attraction could not count as an explanation in mechanical terms. In his *Discours de la cause de la pesanteur* (*Discourse on the Cause of Fall*, 1690) he was content to explain gravity in terms of a refinement of Descartes' vortex theory. To overcome the embarrassing fact that Descartes' original theory could not account for gravitation at the poles of the earth (since the vortex swirls around the equator, at the poles there ought to be no significant centrifugal force, and so no resulting centripetal tendency), Huygens proposed that the particles responsible for gravity circled the earth in *all* directions, around the poles as well as around the equator, and around any other great circle drawn on the surface of the earth (the particles were supposed to be small enough to do so without interfering with one another) [55; 244].

Leibniz, similarly, dismissed Newton's gravity as a 'scholastic occult quality' [110; 121], and remained resolutely unimpressed by Newton's methodological justifications for his concept of gravity. Newton tried to explain forces in terms of the motions of bodies, but those motions, Leibniz insisted, needed to be explained in terms of forces operating upon the bodies; Newton seemed to be letting the tail wag the dog [233; 179].

Ironically, Leibniz's own account of force owed a great deal to the scholastic doctrine of substantial forms, a doctrine which was unanimously decried by every other 'new' philosopher. Leibniz was always keen to reconcile opposing extremes in philosophy and religion and his notion of substantial forms, although very different from the original scholastic notion, can be seen as a way of reconciling scholastic metaphysics with the mechanical philosophy. The scholastic substantial form was what made a given parcel of matter into a specific thing, or individual substance. A chestnut tree and a whale are composed of the same undifferentiated matter, the differences between them are entirely due to their substantial forms. The later scholastics increasingly fell back on substantial forms to explain individual properties of different substances, such as the attractive powers of magnets. This was the kind of useless non-explanation which was widely decried by the new philosophers: it was tantamount to saying that a substance behaved in a particular way because it was in its nature to behave that way [63]. Newton's explanation of gravity, according to Leibniz, was entirely typical of scholastic accounts. But Leibniz himself saw force as a key to making sense of substantial forms within the mechanical philosophy. Cartesian mechanical philosophy endeavoured to expound all phenomena in terms of matter (or extension) in motion, but matter, Leibniz suggested, was nothing more than 'primitive passive force', the ability to resist penetration and movement, while motion was merely a manifestation of changing relationships between bodies as the result of 'primitive active force'. It was this active force, the principle of activity residing in bodies and the cause of their motions, which was fundamental in Leibniz's philosophy [84].

So, the obscure scholastic notions of form and matter were transformed by Leibniz into the mechanical fundamentals of motive force and resistance to motion. This obviously provided a re-affirmation of the mechanicist aim to explain all phenomena in terms of the laws of motion, but it also, to Leibniz at least, provided a metaphysical explanation (or definition) of what force was. While Newton admitted his ignorance of the nature of force, and could claim only to know it by its effects, Leibniz could claim to define it as an essential component of a substance; force not extension was the essence of a body [17; 84; 179].

It should be clear from this that, like Newton, Leibniz went beyond strict mechanistic principles by introducing notions of activity into matter. In Leibniz's case, however, it is easy to see that the original model for his conception of force is based entirely upon the restricted Cartesian notion of force of impact; there is no conception of force capable of acting at a distance in Leibniz's philosophy [233; 3]. Leibniz's fundamental conception of force was summed up in his expression, *vis viva* (living force), which he characterized as a measure of the effect which a moving body can produce, rejecting the Cartesian conception of force as quantity of motion. First published in *Acta Eruditorum* in 1686, Leibniz's concept of *vis viva* went on to stimulate a major controversy which lasted into the eighteenth century [102; 123].

For Leibniz himself it was *vis viva* which was conserved throughout all the physical interactions of the universe, not the quantity of motion, as Descartes had supposed. In developing this notion he rejected the mathematical abstractions of Descartes, Huygens and even Newton, in which colliding bodies were considered to be perfectly hard and to transfer their motions from one to another instantaneously on impact. This could not reflect the reality of the situation, Leibniz insisted, and he proceeded to develop a theory of impacts based upon the assumption of bodies transferring motion from one to another through the compression and subsequent restoration of their supposedly perfectly elastic parts [233; 3; 84].

The transfer of motion in collisions was a *sine qua non* for the mechanical philosophy, being fundamental to its theory of causation. It was also, however, a major stumbling block to those versions of mechanicism which insisted upon the total passivity of matter. It was by no means clear to seventeenth-century thinkers why the motion, but not for example the colour, of a projectile should be transferred upon impact. If matter was entirely inert, why should it respond at all to an impact [78; 89: 390; 233]? Newton circumvented the difficulty by proposing active principles or forces in bodies whose role was, among other things, to take care of inscrutable details like the means of transfer of motion. Leibniz, however, developed a concept of force which was capable of explaining transfer of motion while remaining true to the Cartesian concept of force as force of impact.

There are a number of *prima facie* similarities between Newton and Leibniz: both were profoundly interested in alchemy, and other aspects of Neoplatonic philosophizing, and were to a large extent driven by their own rather idiosyncratic religious beliefs. They were also, of course, consummate mathematicians (indeed, their bitter dispute over priority of invention of infinitesimal calculus clearly intensified their more philosophical and religious differences [97; 3]). More to the point here, they both saw that the only way to extend the mechanical philosophy and to make it really fruitful was to acknowledge that matter itself must be active. As soon as we look beyond these superficial similarities, however, we have to acknowledge the importance of a thorough understanding of their different backgrounds in order to account for the fundamental differences of detail in their natural philosophies.

With the benefit of hindsight it is tempting for the historian to try to make sense of Leibniz's *vis viva* by seeing it as a rudimentary conception of our notion of kinetic energy, but this is liable to distract us from the historians' main aim of understanding the past. Leibniz's concept of *vis viva* was certainly developed within the context of a carefully thought-out metaphysics, which in turn was related to an unshakeable theological position. Leibniz subscribed to an intellectualist theology, which held that there must be absolutes of goodness, justice, and the like, by which even God must abide. Seeing body in the scholastic terms of matter and substantial form, Leibniz proposed that primitive active forces constituted the form of bodies and thereby was committed to a view of bodies in which activity was an inseparable part of their essence. This meant that God could not have created bodies any other way; the concept of passive matter was a contradiction in terms for Leibniz. *Vis viva* and its conservation in all the workings of the universe was essential, therefore, to Leibniz's metaphysics, natural philosophy, and theology [84].

Isaac Newton, by contrast, was a thoroughgoing voluntarist in theology, holding that God's arbitrary will acknowledged no constraints of preconceived absolutes. Whatever God willed was good, by virtue of his willing it. The details of the Creation could only be discovered by experience, not by rational reconstruction of what God 'must' have done. Accordingly, Newton held matter to be active only by virtue of the fact that God had 'superadded' active principles to it at the Creation. Newton was free, therefore, like his God, to conceive of active principles of gravitational attraction, of interparticulate repulsion, of fermentation and other empirical phenomena. Leibniz was constrained by his metaphysics, however, to explain all phenomena in terms, ultimately, of the motive force of *vis viva* [233; 84; 110].

The intermeshing of theology, metaphysics and natural philosophy in the conceptions of both Newton and Leibniz en-

Newton & God

sured that the dispute between them was invested with what might be called cosmic significance. This is most clearly brought out in the exchange of philosophical letters between Leibniz and Samuel Clarke (1675–1729) (acting as Newton's mouth-piece), published shortly after Leibniz's death in 1717. Dealing with the nature of space, time, gravity and force in general, as well as notions of God's interaction with the world, and Leibnizian metaphysical principles like the principle of sufficient reason, the correspondence revealed two irreconcilable world-views [4].

The especial determination of the Newtonians to refuse to concede anything to Leibniz and his followers has been linked to contemporary political developments in England. Leibniz's philosophical theology seemed to English thinkers to show similarities to various radical freethinking political factions, from which the Newtonians wished to dissociate themselves. Their judgements about Leibniz's philosophy and its significance, therefore, were affected by the social and political context of early eighteenth-century Britain [196]. Even the interpretations of later experimental attempts to test the validity of the concept of *vis viva* can be seen to have been affected by the pre-established discord between Newtonians and Leibnizians [123].

The mechanical philosophy was clearly inseparable from developments in the understanding of mechanics, kinetics, and dynamics, but it was also a major force in other aspects of contemporary natural philosophy. The mechanical philosophers sought to show the richness of their new philosophy by showing how it could explain the forms, functions, and vital processes of living creatures. Indeed, it seems fair to say that for two of the leading mechanical philosophers, Descartes and Hobbes, the explanation of vital phenomena and animal (including human) behaviour was always a dominant aspect of their natural philosophies [108; 89; 207].

Descartes' starting point was to draw upon William Harvey's work on the heart and blood, excise it of its vitalistic elements, and, by ignoring Harvey's account of the way the heart moves, produce a mechanistic account of blood circulation. Harvey had demonstrated that, contrary to the ancient Galenic view, the active part of the heart's cycle was its contraction (systole), but he supposed that the activity of the heart was stimulated and maintained by the innate vital power of the blood, endowed with its own pulsific faculty [165; 101: Ch. 17; 75: Ch. 1]. This kind of explanation did not suit Descartes. Adapting earlier notions of an innate, vital heat in living organisms [148], Descartes supposed that there was a fire burning in the left ventricle of the heart. As blood entered the left side of the heart from the cool lungs it was immediately vaporized by the heat, so causing a rapid expansion of the heart and a rapid escape of the vaporized blood out into the aorta, and on through the arterial system. The expanded heart collapsed, just as more blood entered from the lungs to set off the cycle once more [108; 101: Ch. 18; 235: Ch. 5; 148: Ch. 3].

Even though his mechanistic explanation of the heartbeat ran counter to Harvey's elegant experimental demonstrations of systole as the active stroke, Descartes insisted that his explanation of the heart's movements necessarily followed from the disposition of its parts in the same way as the movements of a clock followed from the arrangement of its wheels. The fire in the heart, which is said not to differ from other fires which burn without light in inanimate bodies (he has fermentation in mind – which at that time was not linked to the activity of micro-organisms), is held to be the mainspring and origin of all other bodily movements.

Descartes went on to build a speculative physiology in which animal and human bodies functioned like complex automata based on hydraulic systems. Even though he withheld his *Traité de l'homme* from publication for fear of offending the Roman Catholic Church, he included his account of the movement of the heart and blood in his *Discourse on Method* (1637) as an indication of how a mechanistic physiology would work. It proved to be extremely influential and mechanistic attempts to explain life (or to explain it away?) gathered momentum throughout the century [108; 148; 172].

In England the new mechanistic natural philosophy was allied during the 1650s with the careful anatomical experimentalism inspired by Harvey to give rise to a remarkably coherent research tradition which lasted into the 1670s. The major focus of this research was respiration. The two main speculations about the purpose of respiration – to cool the system down or to

68

introduce air into the left ventricle of the heart for the production of vital spirit – were dismissed in the 1650s. The alternative mechanical hypotheses, that respiration was the means of transporting the blood from right to left ventricle, or of churning and mixing the blood particles, gradually gave way to the suggestion that the blood took a vital ingredient from the air during the transit of the lungs. First put forward in the 1650s, this idea was most forcefully argued by John Mayow (1641–79) in 1668 and 1674, and Thomas Willis in 1670 [75; 101].

The mechanical philosophy was also applied to the understanding of muscular movements, either by means of the mechanical analysis of the loads exerted to move the skeleton (seen as a complex of levers), as exemplified by Giovanni Borelli [235, 101]; or by more speculative theorizing about the chemical means by which muscles were made to contract [75; 101; 235]. Explanations of muscular movement had to be compatible with, and in effect extensions of, explanations of the processing of sensory inputs from outside the body, and explanations of the internal phenomena of appetites and emotions, or 'passions'. Only in this way could all aspects of animal behaviour be reduced to machine-like responses to appropriate stimuli. The mechanical philosophy was used to explain everything which previously had been explained in terms of the operations of the vegetative or animal souls – substantial forms believed to endow living creatures with the powers of reproduction, growth and nutrition (vegetative), and perception, appetites and self-motion (animal). The result was a new concept of living creatures as *bêtes-machines*, which always acted in strict accordance with the laws of mechanics [101; 108].

The major challenge for this aspect of the new philosophy was animal generation. The standard Aristotelian account described the gradual emergence of the differentiated parts of the animal body from an undifferentiated fluid (within the egg or the womb), under the formative action of the male semen or something in the semen. The Cartesian account differed only in so far as he dismissed the Aristotelian assumption that the formative agent was 'soul-like' and possessed intentionality. Instead, Descartes suggested that the mingled semina of the parents fermented, causing an agitation of their particles, which then began to press on other particles, so that they were gradually disposed, in accordance with mechanical laws, to form the parts of the animal fetus. To forestall criticisms about the vagueness of this, Descartes simply asserted that if we knew in sufficient detail the micro-structure of the semen we would be able to deduce the shape and structure of the adult animal 'by reasons entirely mathematical and certain' [241; 172].

Unconvinced, a number of subsequent mechanical philosophers began to develop the concept of pre-existence. One of the most remarkable was inspired by the microscopical anatomical research on insects conducted by Jan Swammerdam. Seeing the rudimentary forms of the nymph within the caterpillar, and the butterfly within the nymph as a result of micro-anatomical dissection (1669), Swammerdam developed the theory of *emboîtement* (1672). There is no metamorphosis of amorphous matter into the organized forms of plants and animals, Swammerdam insisted, only growth into the region of the visible of previously existing invisible parts. The theory of pre-existence is known as *emboîtement* because it envisages all the generations of creatures encapsulated in the eggs of the females: an unborn female will exist already in an egg, and she will have eggs, in some of which will be females with eggs, and so on [241; 101; 172; 74; 1]. There has been a tendency amongst historians to present this theory in crude terms and to dismiss it as ridiculous, but, as recent works have shown, the pre-existence theory in the hands of its chief proponents, Swammerdam and Nicolas Malebranche (1638–1715), was much more subtle [241: 125–8; 172: 244; 182]. In fact, both writers seem to claim only that rudimentary versions of the essential parts of creatures pre-exist in such a way that Cartesian mechanical laws have something upon which they can work. In other words, pre-existence theory should be seen not as a crude picture of animal generations like so many Russian dolls, but merely as a means of saving Descartes' theory of mechanistic epigenesis [182: 222].

The emphasis upon the female egg, which derived primarily from Harvey's research upon the generation of animals (in which he concluded that all living creatures emerged from an egg [241; 101]), was challenged in 1677 by Antoni van Leeuwenhoek, when he discovered spermatozoa. An 'animalculist' alternative to 'ovism', seemed to revert back to the more traditional and long-standing view that males played the most important part in reproduction. Leeuwenhoek himself felt it was more appropriate for the soul to be carried in the male semen, rather than the female egg, and he developed a preformationist theory in which new creatures were elaborated in the male's semen (governed by the father's animal soul), before conception and subsequently grew in the mother's womb with the nutriment provided by the egg [241].

Animalculism never really caught on. The theory suffered a number of drawbacks. Leeuwenhoek himself, as an ill-educated draper, whose empirical results frequently could not be matched by others (his simple microscopes – essentially tiny beads of glass used as extremely powerful magnifying glasses – demanded a careful technique and good eyesight), was often dismissed as unreliable [11; 200: 306–7]. The subject also seems to have raised problems of propriety, involving the public discussion of intimate matters. Leeuwenhoek, for example, felt obliged to insist on one occasion that he had come by his experimental sample of semen not by 'sinful contrivance', but by rushing from the marriage bed to his microscope – but even this smacked of conduct unbecoming [241: 131–2]. Finally, when doubts about the real existence of these animalcules had been overcome, the wider awareness of microparasites made spermatozoa seem unlikely candidates for such a special role in generation. It was assumed they were merely contaminants in the seminal fluid [241: 136–7].

In spite of the seemingly insuperable difficulties of explaining the extreme complexities of living things in mechanistic terms, the mechanical philosophy was as influential in the life sciences as it was in the more physical sciences. It was as though seventeenth-century thinkers refused to acknowledge the difficulties. This is inexplicable if we only consider the technical arguments of mechanistic physiology and anatomy. To understand it we need to consider broader aspects of the intellectual world of the seventeenth century. We must remember, for example, the comprehensiveness of Aristotelianism and the perceived need to replace it with an equally comprehensive system. The perception of that need was inevitably tied in with other aspects of the Aristotelian dominance – its links with the institutions of religion, for example, and other social and cultural institutions. One important aspect of this can be seen in the way educated medical practitioners embraced a mechanistic physiology which enabled them to present a seemingly formidably learned and technical expertise to impress a clientele which had become increasingly disaffected with traditional Galenic medicine [18]. There was even, for a while, a Newtonian version of mechanical physiology, touted by the most ambitious medical practitioners [19; 94]. But perhaps the most obvious aspect of this need for comprehensiveness can still be seen in modern biomedical sciences. Although vitalistic ideas have had their moments in the subsequent history of the life sciences, they have mostly been seen as capitulations to a fundamentally 'unscientific' view, and as such have tended to be reduced, sooner or later, to a more 'mechanist' account. It remains true to say that our own world-view is heavily influenced by the mechanistic notion of the *bête-machine*, with all its implications for biology and medicine. In this sense, the mechanical physiology of Descartes and others can be seen as the origin of the modern biomedical sciences.

5 Religion and Science

There is still a lingering tendency to see science and religion as thoroughly opposed and incompatible approaches to the understanding of fundamental truths about the world. There has been conflict between these two world-views, but that is far from the whole story [15]. Even the so-called 'Galileo affair', probably the most well-known example of scientific knowledge coming into conflict with religion, was by no means the inevitable outcome of two supposedly contradictory perspectives.

Certainly, the Copernican theory was opposed on religious grounds (by Catholics and Protestants alike) from its first appearance [239], but there was no official pronouncement upon it for over 70 years, and then (1616), in spite of the agreement of a group of consultors that heliocentrism was 'formally heretical', the Catholic Church merely suspended its approval of Copernicus' book 'until it should be corrected' [206; 204]. The Church of Rome only really insisted upon the heretical nature of Copernicanism with the condemnation of Galileo in 1633. Historical research has now made it abundantly clear that, far from being the inevitable outcome of a clash between scientific and religious mentalities, the condemnation of Copernicanism and of Galileo was the entirely contingent result of a number of highly specific factors.

The delicate balance which kept Copernicanism away from the serious concern of the Inquisition was disturbed by the talent of Galileo, courtier, for making enemies. He made enemies in the 1610s and 20s of powerful groups of Dominicans, and Jesuits, and a characteristic arrogance displayed in his *Dialogue on the Two Chief World Systems* (1632) succeeded even in alienating his erstwhile supporter, Pope Urban VIII [13; 66; 204; 185; 206]. The situation was not helped by Galileo's insistence upon entering into public discussion of Biblical interpretation (to show how Copernicanism could be made compatible with various Biblical statements), at a time when the Counter-Reforming Catholic Church was trying to restrict free interpretation of Scripture. Furthermore, a series of circumstances in the printing and publication of the *Dialogue* drew (unfounded) suspicion on Galileo as a sympathizer with anti-Papist factions, at a time when Urban VIII was feeling extremely beleaguered [13; 66; 204; 206; 49]. If the outcome was inevitable it was so only because of these highly specific circumstances. The Galileo affair should not be taken as a general indicator of relations between science and religion in the early modern period.

This becomes all the more obvious if we look at almost any other major contributor to the Scientific Revolution. Time and again we can see the importance of religious concerns to the leading thinkers, in providing a general motivation for, and in shaping the precise details of, their natural philosophies. Kepler, for example, saw himself as a priest 'of the Most High God with respect to the Book of Nature' who, by discovering the pattern which God had imposed on the cosmos, was 'thinking God's thoughts after Him' [15: 19–22; 71]. Francis Bacon described his plans for the reform of natural philosophy as a work of preparation for the Sabbath. The Sabbath he had in mind was the ultimate, everlasting Sabbath after the Day of Judgement, which he believed would be ushered in, according to a Biblical prophecy, after the augmentation of the sciences [15: 22; 228; 115]. The natural philosophies of Pierre Gassendi [164], René Descartes [164; 89], Robert Boyle [133; 201; 200; 120], Isaac Newton [57; 236] and Gottfried Wilhelm Leibniz [17; 84; 179] were each carefully developed in order to provide support for the individual theological views of their respective authors. Precisely the same could be said of the natural philosophies of a host of lesser figures from Paracelsus [232] to Blaise Pascal [7; 48], from Joan Baptista van Helmont [133; 166] to William Whiston (1667–1752) [73], from Marin Mersenne [7; 47] to Nicolas Steno (1638–86) [7]. All in all, there can be little doubt of the importance of religious devotion in motivating and shaping early modern science.

One of the major concerns of the mechanical philosophers, for example, was to show how God interacted with the mechanical world. Because of its dependence on quasi-atomist concepts of matter, the mechanical philosophy was easily associated with the supposedly atheistic atomism of the Ancient Greek, Epicurus. According to Epicurus, matter was inherently self-moving, and all things could be explained in terms of the necessary consequences of chance collisions by atoms. Gassendi, who actively sought to rehabilitate Epicureanism for Christian readers, rejected this aspect of Epicurus' matter theory and insisted that God had endowed matter with an internal principle of motion at the Creation [20; 164]. This stratagem was taken up by a number of other mechanists, including Boyle and Newton. In this way God's existence could be proved by pointing to the activity of matter. Since matter was not necessarily active, they supposed, in the way that it was necessarily extended (we can, after all, imagine matter without activity, but we can't imagine it without extension), any activity in matter could be explained only by recourse to God's creative power [110].

Descartes, whose system was based on the assumption that matter was completely passive and inert, had to develop a different stratagem. Since matter was characterized in terms of extension and he wished anyway to avoid suggestions that matter might have intrinsic powers, Descartes turned directly to God to explain the various interactions of matter. God, according to Descartes, not only set the different parts of matter in the world in motion at the creation, but he also maintained the amount of motion in the world, ensuring that motions were transferred from one parcel of matter to another in accordance with the three laws of nature and seven rules of impact. The amount of motion must be conserved and it must always be transferred in accordance with the same laws of nature in order, Descartes supposed, to maintain God's perfect immutability [83; 106; 89: 248–9].

Descartes' views on force, grossly simplified here, continue to cause controversy among specialist historians of philosophy, and, more importantly for us, they were often overlooked by contemporaries. Many believed that Descartes' laws of nature and rules of impact were supposed to be sufficient to explain the workings of the world without any recourse to God, once it was acknowledged that he had set the system rolling. It was all too easy, however, to develop an atheistic version of this account, simply by assuming, as Aristotle had done, that the world was eternal and had always existed the way it does now. If the system had no beginning, God was not necessary at all [116: 173–7].

Perhaps for this reason, a number of Descartes' followers developed the notion of occasionalism, in which God was the only efficient cause at work in the world. The most influential occasionalist was the Oratorian priest Nicolas Malebranche who argued that the laws of nature did not express genuinely causal relationships: when a stone hit a window it was the occasion on which God exercised his causal power, the stone itself had no power to break the window [129: 404–5; 143]. For a number of contemporaries this seemed to make God directly responsible not only for the utterly trivial but also for the downright evil [143].

Leibniz objected to the occasionalist implication that all physics was a perpetual miracle, insisting upon a return to a natural philosophy in which bodies had their own forces by which they could affect things (in accordance with divinely imposed laws of nature). It was evidently important to Leibniz to preserve the transcendence of his God and this necessitated, he believed, making all bodies the source of their own activity. We saw earlier that he revived the scholastic notion of substantial form to enable him to characterize bodies in terms of passive matter combined with primitive active force (see Chapter 4), but he was also convinced of the truth of another scholastic dictum: that only a genuine individual could be self-active. This presented difficulties for the corpuscularist conception of bodies in the mechanical philosophy. Could a body which was composed of atoms or corpuscles be a genuine individual? Considerations like these, together with a number of other metaphysical complexities, led Leibniz to his mature philosophy in which the world was constituted not of atoms but of monads, essentially living creatures with both bodies and souls (and therefore genuine individuals, like human beings) and so capable of being self-active [17; 84; 179].

The nature of force and the activity of bodies (or their lack of activity) was only one aspect of God's relationship with the physical world. Concepts of space were another prime site for discussions of God's place in the world. For Newton, always influenced by Platonic ways of thought, space was 'an emanent effect of God', an outpouring from God's being which provided the immensity of the world. He saw space, therefore, as a real existence and infinite in extent. Indeed, he seems to have gone so far as to later identify space with the immensity of God, so that the Biblical pronouncement that 'In Him we live, and move, and have our being' (Acts 17: 28) was taken quite literally [93]. Newton's concept of absolute space, so important to the elaboration of his *Principia Mathematica* (1687), was not dictated by the requirements of his geometrical analysis of the world system, but by his concept of God.

Leibniz begged to differ. Seeking once again to preserve God's aloof transcendence, Leibniz insisted that space, or dimensionality, could be an attribute of God. If it were, it would mean that God consisted of parts which he took to be absurd. But Leibniz was never content to deny when he could refute. Accordingly, he developed his notion of space as a mere relational concept; an order of co-existences. Extension, shape and motion were only apparent and to a large extent imaginary. It was we, as observers, who imposed extension onto the world. Clearly, therefore, it made no sense to link the absolute God to such a relative space [93].

Natural philosophies which differ fundamentally can often be seen to be grounded in opposed basic assumptions about the nature of God's providence. Voluntarist theology supposes God's will to be his dominant attribute, while intellectualist theology emphasizes God's reason. The voluntarist refuses to acknowledge anything which might circumscribe God's omnipotence, while the intellectualist believes that there are some eternal or pre-existing truths which lead God through his reason to act in certain ways. Voluntarists suppose that what God wills is good, but intellectualists believe that God wills what is good. The voluntarist does not accept that the world can be rationally reconstructed. God's arbitrary will may have introduced any contingency into it, so the system of the world must be discovered empirically. The intellectualist, by contrast, believes that it is possible, at least to a limited extent, to 'think God's thoughts after him' and so arrive at a rational understanding of the world [140; 160; 164].

A number of studies have shown how theological voluntarism or intellectualism has informed the natural philosophies, not only of individual philosophers like Gassendi [164], Descartes

[164], Boyle [133; 160; 200], Newton [57; 236], and Leibniz [140; 160], but also of a whole group of like-minded English thinkers [110; 112]. These studies have shown the powerful interconnections between the underlying theological position on the one hand, and on the other, theories of force and matter, as well as more general epistemological and methodological views. For example, the experimentalism of English natural philosophers, so different from Continental attitudes to experiment, can be seen to dovetail perfectly with the voluntaristic commitment to the unlimited omnipotence of God. Where Descartes, the intellectualist [164], feels he must insist upon the utter passivity of matter, English voluntarists may suppose that God might have endowed matter with intrinsic principles of activity. While Descartes believes that the passivity of matter can be established by the power of reason, English natural philosophers insist upon the experimental investigation of the powers of matter [110; 112].

Another major religious concern of the early mechanical philosophers was the concept of the soul. Each of the first generation of mechanical system builders, Gassendi, Descartes, Sir Kenelm Digby (1603–65), and Walter Charleton, claimed that their mechanical philosophy provided a much better assurance of the immortality of the soul than could Aristotelianism. (Hobbes, missing from this list, rejected the possibility of a disembodied soul.) Their general approach was essentially the same. Having established that all change and dissolution was merely the result of rearrangement or dispersal of the material particles which make up a body, the mechanists could then infer that the rational soul was incapable of change and immortal by virtue of the fact that, being immaterial, it was not composed of material particles [164: 72–3]. It is important to note, however, that this argument only applies to the rational soul which was held to distinguish humans from other creatures. The mechanical philosophers tried to explain the functioning of the so-called vegetative and animal souls in terms of the movements of particularly subtle particles [101; 108; 164: Chs 2 & 9].

Descartes used his mechanical philosophy to underwrite an extreme dualism in which there were two kinds of substance in the world, *res extensa* (an extended thing, or body), and *res cogitans* (a thinking thing, mind or soul). The mind was held to be beyond the bounds of the mechanical philosophy, and Descartes remained essentially silent upon matters which were to tax his followers. How could this immaterial substance cause the body to perform deliberate acts of the will? And whenever it did so, did it not result in an increase in the amount of motion in the world? These were problems for Cartesianism but not, it seems, for Descartes himself [89]. Similarly, Descartes' metaphysical arguments for supposing the existence of a disembodied soul or mind were also fraught with difficulties, being based on little more than our entirely subjective experience that we exist as thinking beings located within a body that is separate, and separable, from ourselves as we really are [89]. The fact that, in spite of these difficulties, Descartes never wavered in his commitment to his dualist system, and always seems to have seen it as a way of demonstrating the immortality of the soul, should suggest to us the influence of religious preoccupations on his thinking. The same could certainly be said of the other mechanical philosophers concerned with the nature of the soul. The differences in detail of the various accounts can always be related to differing religious perspectives [164; 15].

If Descartes' theory of the soul caused internal difficulties for his system, his theory of matter caused difficulties for his Church. The daily miracle of the Eucharist, in which bread was transformed into the body of Christ, was easily explained in Aristotelian terms. Substances (a combination of matter and form) always carried a number of accidental (non-essential) attributes, such as colour, taste and other sensible properties. In the miracle of the Eucharist the accidental properties of the bread remained but the substance was held to have changed. Having done away with this view of body, Descartes attributed properties like colour and taste to the configuration of the particles which made up the body. If bread became flesh it surely had to undergo a change in the configuration of its particles and that must result, by definition, in different sensible properties.

Given the power of the Church, and Descartes' loyalty to his religion, this required some intellectual escapology. Descartes tried two ways of wriggling out of this difficulty. Firstly, he suggested that the surfaces of the bread might remain the same in the Eucharist, thus providing the senses with an unchanged source of sensory information, while the inside turned to flesh. Alternatively, he reverted back to a scholastic explanation in which the substantial form of Christ was said to inform the matter of the bread, in which case, by a scholastic definition, the bread was the body of Christ. For good measure, he suggested a combination of both of these explanations [49].

Descartes was extremely influential and his philosophy quickly gained numerous adherents and even made substantial inroads into the curriculum at a number of universities. The problem of transubstantiation in the Eucharist, however, led to the prohibition of his works by the Congregation of the Index of Forbidden Books in 1663. In 1671 a royal ban on teaching Cartesianism in French universities was issued by Louis XIV. Catholic opposition seems to have been orchestrated to some extent by Jesuits, but in 1678 the Oratorians imposed a ban on teaching Cartesianism in their colleges. Opposition to the philosophy of Descartes was the Catholic Church's most vigorous interference with natural philosophy since the Galileo affair [7]. Unlike Galileo, however, Descartes also invoked the wrath of Protestant authorities. Gisbert Voetius (1588–1676), Calvinist rector of the University of Utrecht in the Netherlands, outraged by some of the Cartesian doctrines of Henricus Regius (1598–1679), campaigned successfully against Cartesianism, as did two theologians at the University of Leiden [89; 129].

Perhaps we should conclude, after all, that science and religion are world-views which are fundamentally at odds with one another? No, again we must resist such a sweeping conclusion. There can be no doubt that his religion was a major stimulus to Descartes' philosophizing and a profound influence upon the details of its development and final form [89; 107]. The same could equally be said of virtually every other leading thinker in the Scientific Revolution. There can be no fundamental incompatibility, therefore, between religious and scientific thought. Nevertheless, major religious institutions, as internally complex and as widely interconnected with other political and social institutions as they are, must respond to a bewildering array of social and intellectual factors. Small wonder that in the politically unstable atmosphere of post-Reformation Europe, religious institutions were sometimes made to act against the burgeoning institution of the new science.

According to a vigorous historiographical tradition, however, there is also a strong case to be made for the *positive* effect on the burgeoning of science by a particular religious institution. It has been suggested that the undeniable success of natural philosophy in seventeenth-century England was due, at least in some large measure, to the rise of Puritanism [151; 33; 227; 228]. This claim has met with considerable opposition, however, and continues to stimulate heated debate.

Part of the difficulty that historians have with the Puritanism-and-science thesis, particularly in the form stated by its principal founder, Robert K. Merton, is that it is hard to see why Puritanism, especially, should have provided a stimulus to science. Suggested factors, such as a concern with socially useful work, 'for the relief of man's estate', as a means of glorifying God and indicating one's state of grace, a concern with rationalism tempered by empiricism [151], rejection of authority in favour of individualistic search for the truth [115], and increased millennial expectations linked to Baconian social ameliorationism [228], can all be seen to be relevant to non-Puritan groups, and in some cases, even to Catholics.

Charles Webster has made the strongest case for links between Puritanism and new attitudes to natural knowledge, agriculture, husbandry, chemistry, medicine, and education during the Civil Wars and the Interregnum [228], but he acknowledges that his study focuses upon a set of rather pragmatic concerns which do not always coincide with present-day notions of what is proto-scientific [228: 517]. Indeed, Webster accepts that the origins of modern science owe more to an ideology alien to Puritanism, but does not say what that ideology is [228: 520]. On the other hand, more recently he has pointed out that a view of 'science' restricted (whiggishly) by present-day concerns is bound to exclude the activities and influence of the Puritan groups which have been at the centre of his research [231: 193]. If, therefore, we accept Webster's definition of science during the Interregnum, we will undoubtedly receive a much fuller understanding of seventeenth-century English attitudes to the natural world than if we choose a definition closer to present-day concerns.

Even so, it is not always possible to see from Webster's work why (or even whether) the reformers he discusses should be regarded as Puritans [158]. But this is to open up another major problem with the Puritanism-and-Science thesis, namely, who was a Puritan and who was not [158]. Merton's original thesis presented the reader with a 'crucial experiment', designed to test the thesis. It consisted of counting the numbers of Puritans in the early Royal Society (founded just after the Restoration of the monarchy in 1660) [151: Ch. 6], but critics were quick to point out that Merton was not an unbiased sampler. Certainly, more careful scholarship has concluded against Merton's view of the early Royal Society [117], and Webster has declared 'headcounting' to be counter-productive in settling the issue [231: 199].

In spite of the difficulties of the Puritanism-and-Science thesis, it still survives as a potent historiographical force. A major reason for this is that opposing views, emphasizing the role of Laudian Anglicanism [218], royalism [157], or a hedonistic-libertarian ethic [70], are even less plausible. Moreover, those alternative views which have seemed plausible seem like mere refinements of Merton's and Webster's views, rather than refutations. In each of these refinements of the Puritanism-and-Science thesis, Latitudinarian Anglicanism plays the dominant role. Barbara Shapiro and others have argued that the sceptical epistemology of the Latitudinarians, which derived from their disgust at the divisiveness of dogmatic pronouncements about the true faith, gave rise among like-minded Anglican natural philosophers to a similar sceptical epistemology in science and a concomitant empirical methodology. The Baconian empiricism, distrust of preconceived theorizing and emphasis on matters of fact which were characteristic of English natural philosophy at this time, can be seen, therefore, to be emulations of the doctrinal minimalism and emphasis on uncontentious matters of faith of the irenic theology of Latitudinarianism [112; 202; 243; 124; 201].

One merit of these claims is that they provide a continuity for the undeniable association between Latitudinarianism and the new philosophy which can be discerned after the Glorious Revolution of 1688. In this later period there was a particular emphasis on the natural philosophy of Isaac Newton and there can be little doubt that the extraordinary pre-eminence of Isaac Newton and his natural philosophy in the cultural life of eighteenth-century Britain owed a great deal to the success of what has been called the 'holy alliance' between that philosophy and the apologetics of low-church Anglicanism [85; 125; 211; 161].

We have been looking at ways in which religion might be said to have promoted the development of natural philosophy and so contributed to the Scientific Revolution (in spite of seeming counter-examples like the Galileo affair and local bans on Descartes' writings on religious grounds). There is another facet to the story, however. There can be no doubt that the late sixteenth and early seventeenth centuries saw not only the origins of modern science, but also the origins of modern atheism. Although it is hardly possible for the historian to point to an out-and-out atheist at this time, there was undoubtedly genuine concern among contemporaries that atheism was becoming increasingly prevalent [119]. It is also clear that the new philosophies were often associated with atheism [234; 119; 15]. It is hardly surprising, therefore, that leading natural philosophers, many of whom, as we've seen, were extremely devout, tried to use their natural philosophies either to defeat atheism, or at least to demonstrate that their philosophies were not atheistic.

This can be seen most clearly in the rise to prominence of the essentially new tradition of physico-theology, or natural theology [15; 234; 37; 175]. Although attempts to prove the existence of God by pointing to the beauty, complexity and order of the natural world – the so-called 'argument from design' – had existed since at least the thirteenth century, it was only in the seventeenth century that whole works of natural history aimed to establish the wisdom and omnipotence of God by scrutinizing the creation. As the new tradition burgeoned, readers were repeatedly told that Nature was God's other book, and that the dedicated student of nature was like a priest.

Natural theology primarily drew upon natural history (see Chapter 2. ii) but natural philosophy soon came in on the act. The latest developments in natural history often derived from studies using the newly invented microscope and these seemed to provide powerful circumstantial evidence for the truth of corpuscularist, and therefore mechanical, natural philosophies [241; 37; 167; 90]. Besides, as we've seen, each of the founders of the new systems of mechanical philosophy specifically intended their respective philosophies to provide an underpinning to religion. If a new philosophy was intended to entirely replace the traditional, comprehensive Aristotelianism, it had to be seen to be capable of taking over the role of handmaiden to religion. Gassendi, accordingly, took great pains to 'baptize' Epicurus, most notorious of ancient atheists [164].

Concern about the increasing prevalence of atheism ran particularly high in England during the Interregnum and the Restoration, and leading natural philosophers often took an apologetic and highly defensive line in the presentation of their natural philosophies [119; 234; 116: Ch. 7]. Even Robert Boyle, whose orthodoxy was never in question, felt obliged to defend his promotion of the new philosophy against charges that it was inherently atheistic [119; 187; 120]. Boyle was so concerned about this that he established just before his death in 1691 an annual series of monthly lectures (the Boyle Lectures) to combat infidelity of various sorts, but first and foremost, atheism [125; 211].

For the most part the defence of the new philosophies against charges of atheism concentrated on the argument from design, pointing to the beauty and complexity of nature, and trying to demonstrate the impossibility of such intricacies without the creative intervention of a supreme artificer [15; 37; 90; 175; 241; 234]. But there were one or two less predictable developments. Drawing upon the still highly respected humanist tradition of historical scholarship, for example, a number of natural philosophers tried to establish that atomism was not an invention of pagan Greek philosophy but a more ancient philosophy in the Judaeo-Christian tradition. A Phoenician by the name of Mochus was increasingly discussed in works of classical scholarship as the founder of atomism, and on at least one occasion he was identified with Moses himself [184]. This tradition was repeatedly invoked in defence of atomism by English natural philosophers throughout the seventeenth century.

Less widespread, but certainly noticeable in the historical record, was the attempt to affiliate the new philosophy to belief in witchcraft and demons. To deny the Devil and all his works was also to deny God. A minority among the natural philosophers, in their concern to dispel charges of atheism, discussed famous and not so famous cases of witchcraft, hauntings and all other evidence of a supposedly spiritual realm. As with arguments to prove the immortality of the soul (see above), the spirituality of these phenomena could be established by showing that they could not be explained in terms of the mechanical philosophy. By showing the reality of the spiritual world, the benefit to religion of the mechanical philosophy was conveniently displayed [119; 128; 230; 188; 203: Ch. 6; 201]. Additionally, by defining all the correctly attributable forms of physical causation in the world, the mechanical philosophy made it easier to determine which explanations were illegitimate. Anyone who merely believed in the efficacy of causal links which mechanical philosophers rejected as unworkable could be held to be misled, either by their own superstition or by the Devil, and therefore guilty (in either case) of turning away from God. The mechanical philosophy thus showed itself to be useful in the war against irreligion [26; 25].

The various apologetic stratagems developed by English natural philosophers as a result of contemporary fears of atheism cannot entirely be separated from the more positive religious intentions which thinkers like Bacon, Boyle, Newton and a host of others cultivated. The result is a subtle and complex set of interactions, often differing from one individual to another [119]. It is important, therefore, to be aware of the interplay of evangelism and apologetics in the natural theologies of seventeenth-century natural philosophers [15]. There can be no doubt, however, that religion and theology played a major part in the development of modern science.

6 Science and the Wider Culture

Throughout this brief survey we have noted the cultural and social context which is so often necessary to our understanding of developments in science. From wider cultural influences such as religion, and the magical world-view, to more specific aspects of social organization, such as those which form the background to developments in the status of mathematical [12; 238] or medical [35; 174; 18; 19; 94] practitioners, from the links between views of God's relationship to the world, correct forms of kingship and legitimate forms of scientific method [196; 140; 160; 124], to the newly perceived need for the pragmatic innovations of elite craftsmen as a background to experimentalism [9; 180; 225; 245], we have seen how developments in early modern science are aspects of changes in the wider culture. In this final chapter we will look at a number of other topics which have been seen as important background elements in the Scientific Revolution.

The period of the Scientific Revolution coincides to a large extent with the beginnings of modern capitalism. The Puritanism-and-Science thesis, in its original formulation by Robert K. Merton, was partly inspired by the earlier work of the sociologist Max Weber, which linked the supposed Protestant work-ethic with the 'Spirit of Capitalism'. Accordingly, Merton was concerned not just with religious beliefs but with concomitant social factors such as the rise of the bourgeoisie, the origins of capitalism and the move towards political reform [151]. In seeing economic stimuli towards improvement of mining techniques, and associated technologies concerned with introduction of fresh air and removal of water, improvements in transportation, navigation, and various military innovations, Merton was adding his voice to a number of other social and economic historians, writing in the 1930s [95: Ch. 1]. Since then a number of other historians have tried to forge links between capitalism and aspects of the new science [211; 125; 124; 227, 150].

Some of this work has pointed to the emphasis on practical utilitarian concerns in the influential writings of Francis Bacon [67; 180]. But a recent study of Bacon has focused more specifically upon his belief that natural philosophy should be capable of providing support for the imperial state [145]. For Bacon natural philosophy should not be an ivory-tower pastime for recluses, but a major collaborative effort for the good of the 'commonweal', 'a kind of royal work' carried out effectively by a department of state with its own royal governor [145: 163]. It is highly significant that one of the clearest statements of how Bacon imagined this royal work should be carried out appears in his fable of 'Salomon's House', a government research institution in the imperial state of Bensalem, imagined in his *New Atlantis* (1627) [145: 135–40].

Commentators upon Bacon have often been puzzled by the idiosyncratic nature of his experimentalism, principally because his notion of experiment does not seem to conform to our own. We can now see that this is principally because Bacon believed that nature could be investigated by the same method as a lawsuit in a courtroom trial [145: 164–71; 150: 168–9; 203: 169]. The analogy between the workings of the law and the investigation of nature has also been seen in subsequent English natural philosophy. The method of Robert Boyle, for example, has been seen as a Baconian enterprise, modelled on the method of English common law, in which 'moral certainty' about physical matters can be arrived at by bringing to bear specific, local experiences, background knowledge, skill, expertise and reason [187: Ch. 2; 201: Ch. 2].

Similarly, the importance of public witnessing of experimental results, emphasized by Boyle and other fellows of the Royal Society as a guarantee of the reliability of the Society's pronouncements, is based upon the authority of legal procedures [201: Ch. 2]. One of the duties of the jurors in a trial, however, was to decide upon the reliability of witnesses. It was taken for granted that some witnesses were more likely to be truthful than others. Here again, the relevance of considerations like these to the new philosophy has been demonstrated by pointing to the gentlemanly ethos of English natural philosophy in the Restoration. Gentlemen were for a variety of reasons the most reliable and truthful witnesses of natural phenomena, and the reputation of the Royal Society owed much to its image, carefully fostered and preserved, as a gathering of gentlemen [200; 201].

For Bacon, then, problems of knowledge, that is to say the problem of how best to arrive at truth and the problem of convincing all onlookers that it is truth, are properly part of a statesman's concerns [145: 141]. Similarly, it has been claimed that for Boyle and other leading members of the Royal Society, solutions to the problems of knowledge were seen as solutions to the problem of how to establish and maintain order in the state [201: 332]. The reliable witnessing of experiments by gentlemen was the only sure way to establish matters of fact about the physical realm. The matters of fact could then be said to have been established with no reasonable possibility for dissent. For Boyle and like-minded thinkers in the Royal Society, reliable natural philosophy should be confined to the establishment of matters of fact. Theorizing and hypotheses were, rhetorically if not actually, eschewed. This solution to the problem of dissension in natural philosophy could be the model for avoiding dissent in religion and polity which had so disrupted affairs in England's recent past, and which continued to threaten the Restoration [201: Chs 7 & 8].

It should be seen, therefore, that there is a strong case to be made for the influence of socio-political considerations on the development of the experimental method in seventeenth-century England. The political situation in England was, of course, unique and highly specific – no other European country experienced anything like the rebellion leading to the Civil Wars, followed by the political instabilities of the Interregnum, and the tensions of the early Restoration. Similarly, the experimental method in England, as we have already seen (see Chapter 2. ii), developed very differently from the way that it did abroad. Recent attempts to show the actual effect of the religious and political background on the development of the experimental method show that the uniqueness of the English in these two spheres is not simply coincidence [48; 112; 202; 243].

There is also some highly interesting work which argues for the impact of political developments on the actual intellectual content of English science. We have already noted the relevance of religious and political reform and counter-reform to the fortunes of Paracelsianism in mid-century England [149; 174; 178]. But the Puritan Revolution has also been seen as a factor in the marked shift in William Harvey's presentation of his ideas on the heart and blood, from an emphasis on the primacy of the heart in 1628 to an emphasis on the blood in 1649. The suggestion here is not that Harvey changed from a monarchist to a Republican in the intervening period, but that he was sufficiently affected by political developments to represent, perhaps even to see, the natural world in a different way. In 1628 Harvey dedicated his *Disputation on the Motion of the Heart and Blood* to Charles I by drawing upon the age-old parallel between the heart and the king [113: 160]. By 1649, the year of the execution of Charles, the heart is described by Harvey in entirely functional terms. Instead of writing of the sovereignty of the heart, Harvey now talks of 'the prerogative and antiquity of the blood': 'the blood lives and is nourished of it selfe, no way depending upon any other part of the body, as elder or worthier than itself' [113: 162]. It would seem that Harvey saw the workings of the heart and blood by analogy with absolutist monarchy in 1628, but by 1649 he could see the system in terms closer to the contract theories of monarchy being developed, for example, by his friend and admirer, Thomas Hobbes. The heart now served the blood, as the king served his people.

Could such a meticulous and careful experimenter as William Harvey really have been so swayed by political concerns? He does, after all, explain in *De Circulatione Sanguinis* (1649) and in *De Generatione Animalium* (1651), the observational and experimental grounds for his belief in the primacy of the blood. Certainly he does, but in so doing he does not live up to the reputation that has been bestowed upon him by his modern hagiographers. On this issue, modern commentators part company with Harvey: he claims here to be seeing something that isn't there, or goes further than the actual observations warrant. Why? Perhaps because he is seeing the body in terms of the body politic. Remember, we are looking back to a time when the boundaries between religion, politics, and philosophy were not so clearly demarcated. God created the natural world and the social world, the same order of rank above rank seemed to pre-modern thinkers to appear in both, and the perceived fact that the political realm, when properly organized, reflected the natural, was routinely taken as proof that all was well in the political realm. What we might regard as mere metaphors were taken, in the early modern period, to reflect the real nature of God's creation [140; 160; 217].

There can be no doubt, for example, that political symbolism was routinely attached to cosmology and correct interpretation of the symbolism was frequently argued in political discourse [122]. The Ptolemaic arrangement of the heavenly bodies located the Sun, most common symbol of the king, among the planets, so suggesting that the king and the nobility (represented by the planets) shared political authority, with much of the king's power mediated by the nobles. The Copernican scheme, however, was seen to lend itself much more easily to the support of more absolutist forms of monarchy. As monarchs laid increasing claim to absolute rule, diminishing the power of local gentry, the Copernican cosmology became increasingly useful. It should not be supposed that it merely suited absolutists to be able to present Copernican cosmology as an image of the new order: it was virtually essential for them to be able to point to a *natural* support for their political claims. Only in this way could they satisfy the general expectation that, in God's creation, the order of the cosmos would, in a fairly obvious way, reflect the order of society. But this should not be taken as a claim that Copernicus and his followers deliberately developed their astronomy in order to further their political beliefs. The increasing acceptance of Copernican heliocentrism, however, can perhaps be taken as an indication of the fundamental shift in ways of seeing what was taken to be the *natural* order of things.

Similar arguments, hinging upon the ideological power of metaphor, have even been used to explain why self-regulating or feedback devices made no appearance in medieval or early modern technology. A number of feedback devices were described in the *Pneumatics* of the Hellenistic Greek writer, Hero of Alexandria (*fl.* AD 1st century) which was first printed in

1575 and went on to become immensely influential, and yet none of those feedback devices were taken up, or modified to appear in any other form [146: xv–xvi]. Feedback devices were simply disregarded throughout Europe until English technicians and engineers began to develop them in the eighteenth century. Even then, the rest of Europe took some time to catch up. Why was this?

In order to understand this aspect of the history of technology we must consider the importance of the so-called clock metaphor in early modern European culture. From the invention of the mechanical clock at the end of the thirteenth century, the clock increasingly came to be seen as a metaphor for the order and regularity of the world. God was seen as a clockmaker and the workings of the clock displayed the importance of fulfilling one's allotted role and obeying the authority of the system; by association, this became a metaphor for the effectiveness of absolute monarchy [146: Chs 2, 4 & 5]. The metaphor frequently appeared, of course, in the writings of the mechanical philosophers: used as a convenient way of illustrating their new philosophy, it served to provide support to the mechanical world-view while simultaneously reinforcing the wider appeal of the metaphor itself.

In Britain, however, particularly towards the end of the seventeenth century, the clock metaphor was treated with a good deal more reserve and ambivalence than on the Continent. The clock often appeared here as a metaphor for regimentation, and mindless compulsion [146: 123–6, 99–101]. It seems fair to say that the metaphor was largely rejected in Britain for the same reason that it was embraced upon the Continent: as a symbol of absolute authority. Contrasting attitudes to the clock metaphor, therefore, reflected different conceptions of order: authoritarian on the Continent and liberal in Britain.

Britain became the leader in European clockmaking in the second half of the seventeenth century but British clocks can be seen to be merely functional in comparison with the grandiose and elaborate clocks of Continental manufacture, which often included automata depicting the movements of the heavens and other elaborations of the cosmic metaphor [146: Ch. 1]. It is as though British clockmakers belittled clocks, as their compatriots cut the clock metaphor down to size. Subsequently,

British engineers were the first to develop self-effacing self-regulating devices (entirely different from the more spectacular technology of clockwork) for use in various kinds of machines. Here again, however, we can perhaps discern the unconscious inspiration of unexamined presumptions about correct social order, since British political and economic theory, after the Glorious Revolution, was frequently couched in terms of 'checks and balances', equilibrium, and other self-regulatory locutions [146: Chs 7, 8, 9 & 10].

If clockwork provided a new metaphor for the cosmos, society and natural philosophy, so too, it has been argued, did a new emphasis on the subjection and control of women. A number of feminist historians and philosophers have pointed to the use of sexual metaphors in the Scientific Revolution to exemplify and justify the new approach to nature [150; 24; 132]. Bacon, for example, spoke of Nature, as though a female, being 'bound into service', put 'in constraint' and enslaved to the natural philosopher. It is no good to clutch at her without laying hold of her, he wrote, Nature must be captured and her secrets, like her inner chambers, penetrated [150: 169–70; 132: 36]. Robert Boyle, similarly, spoke of natural philosophers desiring to command nature, to make her 'serviceable to their particular ends, whether of health, or riches, or sensual delight' [150: 189]. Based on a historiography which sees the magical and scholastic world views as holistic and vitalistic and regards premechanistic views of nature as predominantly feminine in ethos, this work characterizes the mechanical world-view as manipulative, exploitative, and masculine. The mechanical philosophy provided an answer to the problem of cosmic and, therefore, social order, but in so doing it pointed to the need for power and dominion over nature [150: 215].

It is not being suggested that part of the reason for developing the mechanical philosophy was to subjugate women, nor is it claimed that its alleged anti-feminism was part of the reason for its success. The realpolitik of sexual domination surely had no need for help from natural philosophy. But the sexual metaphors which occurred to the new natural philosophers reflected, and helped to shape, attitudes to legitimate knowledge and appropriate knowledge-producers which remain gendered to this day [132; 24; 190; 200: 86–91].

Women have played no part in this brief outline of the Scientific Revolution. Not because there were no women participants, but because their presence was so small and so hardly felt that it seemed inappropriate to include them in such a brief survey. Only Anne, Lady Conway (1631–79), who has been claimed as an influence on Leibniz [150, 5], and Emilie du Châtelet (1706–49), whose *Institutions de physique* (1740) and French translation of Newton's *Principia* (1759) helped to introduce Leibniz's and Newton's work to France [190; 213], have any real claims to a significant contribution to the Scientific Revolution. Margaret Cavendish (1623–73), Duchess of Newcastle, published a number of highly idiosyncratic and fascinating works of natural philosophy but seems to have met only with indifference or ridicule [190; 5; 186]. It seems true to say, therefore, that the main contribution to history of these, and a number of other females less well-connected to the gentlemanly circles which predominantly produced the new science, lies in demonstrating to historians what women were capable of, in spite of the almost insurmountable barriers which their society erected against them [5; 190; 213].

There are other aspects to the socio-political dimension of the Scientific Revolution, and no doubt more will be discerned in future research. But one general point can already be made with confidence and that is that it will not do to suggest that while natural philosophy can be put to use for political purposes, the natural philosophy itself is somehow pre-social and 'purely intellectual' [195]. We have seen here that a number of aspects of the social structure are crucially relevant to our understanding of the origins of modern science.

This, finally, leads us to the even more general conclusion that if we wish to achieve as full an understanding as possible of the Scientific Revolution we need to consider not only the role of religion, theology, politics, economics, metaphysics, methodology, and technical issues but also the complex interplay between all these factors. Only by means of such a rich synthesis can we hope to understand the cultural phenomenon which has been seen as the 'the real origin both of the modern world and of the modern mentality', and whose significance has been said to reduce the Renaissance and the Reformation 'to the rank of mere episodes' [21: vii].

7 Conclusion

Writing in the 1720s, Bernard Le Bovier de Fontenelle (1657–1757), permanent secretary of the Académie Royale des Sciences from 1699, pointed to the development of infinitesimal calculus by Newton, Leibniz, Jacques Bernoulli (1654–1705), Pierre Varignon (1654–1722) and others and called it 'an epoch of almost total revolution occurring in geometry' [32: 212; 21: Ch. 9]. In the 1750s the two editors of the *Encyclopédie*, Denis Diderot (1713–84) and Jean Le Rond d'Alembert (1717–83), talked of the revolution in science which had been initiated in the previous century and which they saw as continuing [32: 217–20; 103: 1–2].

The major inspiration behind such eighteenth-century perceptions of revolutionary change in science was undoubtedly Isaac Newton. French intellectuals, for a while unimpressed, began to acknowledge Newton's superiority to Descartes after the appearance of Voltaire's (1694–1778) *Letters on the English Nation* (1734) and *Elements of Newton's Philosophy* (1738), and after Maupertuis (1698–1759) published the results of his attempt to ascertain the shape of the earth (1739), which had been recognized as a test-case for Cartesian and Newtonian theories (if Descartes was right the earth ought to look like an egg standing up tall and spinning on its end, if Newton was right it ought to look more like an egg spinning on its side). Declared, by D'Alembert in the 'Preliminary Discourse' to the *Encyclopédie* (1751), to be a great genius who finally established the correct form of natural philosophy, Newton's influence in the eighteenth-century and beyond was unparalleled [99: Ch. 14; 91: Ch. 5; 103; 109; 125; 189]. His *Principia Mathematica* was widely regarded as the very model of the new mathematical way of doing physics, while the *Opticks* was taken as a model of experimentalism. Both of his alternative speculations about the causes of natural phenomena – attractive and repulsive forces between the particles of bodies, or an all-pervasive subtle aether constituted of repelling particles – proved influential in the development of chemistry and theories of electricity, heat and light [103; 109; 189]. The perceived success of Newton's method also meant that it was frequently invoked in Enlightenment attempts to develop a 'science of man', embracing sensationalist psychology, civic morality and political economy [91; 102; 161].

It would be wrong, however, to reduce the legacy of the Scientific Revolution to the legacy of Isaac Newton, and not just because, as he himself admitted, he saw further only by standing on the shoulders of giants. Eighteenth-century developments in mathematics perhaps owe more to the achievements of Leibniz and the Bernoulli brothers, than to Newton, whose dominion over British mathematicians seems to have led to a noticeable decline (usually attributed to the awkwardness of Newton's system of notation compared to that of the Continental mathematicians) [99; 103; 91]. Newton did not have it all his own way in physics either. The *vis viva* controversy, for example, revealed that Newton had not had the full measure of force [99; 103; cf. also 23]. Although Newton's belief in attractive and repulsive forces in matter had an influence in chemical theorizing, it was only one strand in a complex network of earlier theories and practices deployed by eighteenth-century chemists [103]. If Newton's methodology was invoked to support the claims of the newly conceived social sciences, the content owed more to earlier political and moral theorists like William Petty (1623–87), Thomas Hobbes, and John Locke [161]. Although there was a vogue for medical Newtonianism at the beginning of the eighteenth century it proved to be short lived and, for the most part, the bio-medical sciences proceeded along lines laid down before Newton's name became so potent [19; 94]. There was also no shortage of opposition to Newton and his philosophy [23].

In order to comprehend Newton's towering presence in the Enlightenment we should bear in mind that the earliest claims that a revolution in science had taken place were themselves aimed at bolstering the intellectual authority of natural philosophy. Men like Fontenelle, the *Encyclopédistes*, Voltaire and other Enlightenment *philosophes* had their own reasons for wanting to present natural philosophy as a newly powerful and reliable system of knowledge, pointing the way to progress and the improvement of the human condition. In need of hero figures to represent the strengths of this movement, these writers turned to Descartes to represent rationalism, to Bacon to represent experientialism, and above all to Newton, who represented the triumphant synthesis of both methods [32; cf. 50].

It was not merely convenience, much less coincidence, which led Enlightenment intellectuals to see natural philosophy as a means of promoting their own belief in the authority of reason and experience, and in the force and reliability of naturalistic arguments. They were after all the immediate heirs to the radical changes in intellectual life which had been brought about by the period which they began to see as one of Scientific Revolution. In the end, therefore, it is possible to conclude that the very fact that they now saw natural philosophy in this way, and even dared to hope that it might be used to establish laws for the correct ordering and running of society, is in itself indicative that a revolution in the ordering of knowledge had indeed taken place. The Scientific Revolution was complete.

Bibliography

With one or two exceptions, the following list confines itself to what are called secondary sources – sources written by historians about the past. For a proper understanding of past views of the natural world there is, however, no substitute for reading the primary sources – texts written by the scientists themselves. The main works of all the major figures in the history of science are easily available in English translation and should certainly be consulted if you wish to get any real sense of their concerns and methods. In some cases there are also English editions of manuscript writings and correspondence which provide major sources for historical understanding.

In addition to the secondary sources mentioned here there are a number of other useful guides. There are significant articles on all the scientists mentioned here, together with bibliographies, in C. C. Gillispie (ed.), *Dictionary of Scientific Biography*, 16 volumes (New York: Scribners, 1970–80). For preliminary indications about various topics, consult W. F. Bynum, E. J. Browne and R. Porter (eds), *Dictionary of the History of Science* (London: Macmillan, 1981). Useful literature surveys (up to their date of publication!) are provided in P. Corsi and P. Weindling, *Information Sources in the History of Science and Medicine* (London: Butterworth Scientific, 1983); and P. T. Durbin (ed.), *A Guide to the Culture of Science, Technology, and Medicine* (New York: Free Press, 1980). Those interested in the historiography of the Scientific Revolution should consult H. Floris Cohen, *The Scientific Revolution: A Historiographical Inquiry* (Chicago: University of Chicago Press, 1994).

[1] H. B. Adelman, *Marcello Malpighi and the Evolution of Biology*, 5 vols (Ithaca: Cornell University Press, 1966). Massive edition of Malpighi's works which includes an intellectual biography (v. 1) and

a study of the history of embryology (v. 2).

[2] E. J. Aiton, *The Vortex Theory of Planetary Motions* (London: Macdonald, 1972). Classic account of Cartesian cosmology.

[3] E. J. Aiton, *Leibniz – A Biography* (Bristol: Adam Hilger, 1989). Fullest account of Leibniz's life and work, but needs to be supplemented by more analytical studies [17, 84, 179].

[4] H. G. Alexander (ed.), *The Leibniz Clarke Correspondence* (Manchester: Manchester University Press, 1956). Important exchange of letters between Leibniz and Newton's mouthpiece, Samuel Clarke, with a useful introduction by the editor.

[5] Margaret Alic, *Hypatia's Heritage: A History of Women in Science from Antiquity through the Nineteenth Century* (London: Women's Press, 1986). A survey of women's involvement in the history of science.

[6] Agnes Arber, *Herbals: Their Origin and Evolution. A Chapter in the History of Botany* (Cambridge: Cambridge University Press (1912), 3rd edn, with Introduction and Notes by W. T. Stearn, 1986). Still the best survey of herbals and their significance.

[7] William B. Ashworth, Jr, 'Catholicism and Early Modern Science', in D. C. Lindberg and R. Numbers (eds), *God and Nature: Historical Essays on the Encounter between Christianity and Science* (Berkeley: University of California Press, 1986), pp. 136–65. Excellent brief survey of Catholics and science after Galileo.

[8] William B. Ashworth, Jr, 'Natural History and the Emblematic World View', in David C. Lindberg and Robert S. Westman, *Reappraisals of the Scientific Revolution* (Cambridge: Cambridge University Press, 1980), pp. 303–32. Important article on the history and significance of natural history before and during the Scientific Revolution.

[9] J. A. Bennett, 'The Mechanics' Philosophy and the Mechanical Philosophy', *History of Science*, 24 (1986), 1–28. Revisionist essay which argues for the role of mathematical practitioners in the establishment of the experimental method and the mechanical philosophy.

[10] J. A. Bennett, 'The Challenge of Practical Mathematics', in S. Pumfrey, P. Rossi and M. Slawinski (eds), *Science, Culture and Popular Belief in Renaissance Europe* (Manchester: Manchester University Press, 1991), pp. 176–90. A useful supplement to [9].

[11] Klaas van Berkel, 'Intellectuals against Leeuwenhoek: Controversies about the Methods and Styles of a Self-Taught Scientist', in [167], pp. 187–209. Fascinating account of contemporary reactions to Leeuwenhoek and his work. Shows role of social factors in intellectual life.

[12] Mario Biagioli, 'The Social Status of Italian Mathematicians, 1450–1600', *History of Science*, 27 (1989), 41–95. Excellent preliminary survey of factors involved in changing status of mathematical practitioners. Includes a bibliography three times bigger than this one.

[13] Mario Biagioli, *Galileo, Courtier: The Practice of Science in the Culture of Absolutism* (Chicago: Chicago University Press, 1993). Lively account of Galileo's social setting, showing its influence on the actual content

of his science. Perhaps exaggerates the extent to which Galileo played the courtier's role.

[14] J. Bruce Brackenridge, 'Kepler, Elliptical Orbits, and Celestial Circularity: A Study in the Persistence of Metaphysical Commitment', *Annals of Science*, 39 (1982), 117–43, and 265–95. Detailed but compressed account of Kepler and the harmony of the spheres.

[15] John Hedley Brooke, *Science and Religion: Some Historical Perspectives* (Cambridge: Cambridge University Press, 1991). Fullest and most nuanced survey of the interactions between science and religion.

[16] Harcourt Brown, *Scientific Organizations in Seventeenth-Century France* (Baltimore: Johns Hopkins University Press, 1934). An early but still useful study of the institutionalization of science.

[17] Stuart Brown, *Leibniz* (Hassocks: Harvester Press, 1984). Helpful survey of Leibniz's major philosophical concerns.

[18] Theodore M. Brown, 'The College of Physicians and the Acceptance of Iatromechanism in England, 1665–1695', *Bulletin of the History of Medicine*, 44 (1970), 12–30. Explains why the mechanical philosophy became important to the professional aspirations of medical practitioners.

[19] Theodore M. Brown, 'From Mechanism to Vitalism in Eighteenth-Century English Physiology', *Journal of the History of Biology*, 7 (1974), 179–216. Details the intellectual and social reasons for the decline of mechanism in medical theory.

[20] Barry Brundell, *Pierre Gassendi: From Aristotelianism to a New Natural Philosophy* (Dordrecht: D. Reidel, 1987). Detailed account of how Gassendi developed atomism to replace all aspects of Aristotelianism.

[21] Herbert Butterfield, *The Origins of Modern Science, 1300–1800*. 2nd edn (London: Bell, 1957). A classic study, many times superseded but still very useful, and not just for understanding the preoccupations of subsequent historians of science.

[22] Jerome Bylebyl, 'The School of Padua: Humanistic Medicine in the Sixteenth Century', in C. Webster (ed.), *Health, Medicine and Mortality in the Sixteenth Century* (Cambridge: Cambridge University Press, 1979), pp. 156–69. Good account of the intellectual and methodological importance of anatomy in Europe's leading medical school.

[23] Geoffrey Cantor, 'Anti-Newton', in J. Fauvel, R. Flood, M. Shortland and R. Wilson (eds), *Let Newton Be!* (Oxford: Oxford University Press, 1988), pp. 203–21. Compressed but indicative survey of opposition to Newton, and the reasons for it, in the eighteenth century.

[24] J. R. R. Christie, 'Feminism and the History of Science', in R. C. Olby et al. (eds), *Companion to the History of Modern Science* (London: Routledge, 1990), pp. 100–9. Brief survey of feminist approaches to history of science. Includes further readings.

[25] Stuart Clark, 'The Scientific Status of Demonology', in [222], pp. 351–74. Offers surprising and important insights on the nature of early modern magical theory.

[26] Stuart Clark, 'The Rational Witchfinder: Conscience, Demonological

Naturalism and Popular Superstition', in S. Pumfrey, P. Rossi and M. Slawinski (eds), *Science, Culture and Popular Belief in Renaissance Europe* (Manchester: Manchester University Press, 1991), pp. 222–48. Important account of the social and religious background to early modern attempts to define natural causality.

[27]   Antonio Clericuzio, 'A Redefinition of Boyle's Chemistry and Corpuscular Philosophy', *Annals of Science*, 47 (1990), 561–89. Important article showing that Boyle's matter theory cannot be understood in mechanistic terms.

[28]   Antonio Clericuzio, 'From van Helmont to Boyle: A study of the Transmission of Helmontian Chemical and Medical Theories in Seventeenth-Century England', *BJHS*, 26 (1993), 303–34. Useful study of the role of chemical theorizing in the development of the new philosophy in England.

[29]   Nicholas H. Clulee, *John Dee's Natural Philosophy: Between Science and Religion* (London: Routledge, 1988). Best account of John Dee and an excellent study of the role of magic in the making of modern science.

[30]   V. Coelho (ed.), *Music and Science in the Age of Galileo* (Dordrecht: Kluwer Academic, 1992). Advanced collection of essays showing the close affiliation between music and natural philosophy.

[31]   I. B. Cohen, 'The *Principia*, Universal Gravitation, and the "Newtonian Style", in Relation to the Newtonian Revolution in Science', in Z. Bechler (ed.), *Contemporary Newtonian Research* (Dordrecht: Reidel, 1982), pp. 21–108. Presents Newton's methodology as a distinctive and revolutionary way of doing natural philosophy.

[32]   I. B. Cohen, *Revolutions in Science* (Cambridge, Mass.: Harvard University Press, 1985). A history of the historians' concept of scientific revolutions. Full of fascinating material.

[33]   I. B. Cohen (ed.), *Puritanism and the Rise of Modern Science: The Merton Thesis* (New Brunswick: Rutgers University Press, 1990). Useful compilation of articles for and against the Puritanism-and-Science thesis, with a bibliography.

[34]   H. Floris Cohen, *Quantifying Music: The Science of Music at the First Stage of the Scientific Revolution, 1580–1650* (Dordrecht: Reidel, 1984). A full survey of the musical theories of leading natural philosophers, and their significance.

[35]   Harold J. Cook, *The Decline of the Old Medical Regime in Stuart London* (Ithaca, NY: Cornell University Press, 1986). Excellent history of the English Royal College of Physicians which also offers wider insights on the relationship between medicine and natural philosophy.

[36]   Harold J. Cook, 'The New Philosophy in the Low Countries', in R. Porter and M. Teich (eds), *The Scientific Revolution in National Context* (Cambridge: Cambridge University Press, 1992), pp. 115–49. A good survey of science in the Netherlands in a generally very useful anthology.

[37]   Harold J. Cook, 'The Cutting Edge of a Revolution? Medicine and

Natural History near the Shores of the North Sea', in J. V. Field and
F. A. J. L. James, *Renaissance and Revolution: Humanists, Scholars, Crafts-*
men and Natural Philosophers in Early Modern Europe (Cambridge: Cambridge University Press, 1993), pp. 45–61. A well-argued plea to
remember the importance of non-mathematical sciences in the Scientific Revolution.

[38] Brian Copenhaver, 'Astrology and Magic', in C. B. Schmitt and Q.
Skinner (eds), *The Cambridge History of Renaissance Philosophy* (Cambridge:
Cambridge University Press, 1988), pp. 264–300. Excellent but advanced
brief survey.

[39] Brian Copenhaver, 'Natural magic, Hermetism, and Occultism in Early
Modern Science', in D. C. Lindberg and R. S. Westman (eds), *Reap-*
praisals of the Scientific Revolution (Cambridge: Cambridge University
Press, 1990), pp. 261–302. Seeks to clarify and explain differences
between magic, the occult and hermeticism.

[40] Lesley B. Cormack, 'Twisting the Lion's Tail: Practice and Theory at
the Court of Henry Prince of Wales', in [155], pp. 67–83. A case-study
clearly illustrating the role of patronage in natural philosophy.

[41] A. C. Crombie, *Robert Grosseteste and the Origins of Modern Science* (Oxford: Clarendon Press, 1953). An important study of a major figure in
medieval science which makes claims about the origins of
experimentalism before the Scientific Revolution.

[42] A. C. Crombie, *Augustine to Galileo,* 2 vols (London: Heinemann, 1959).
Major statement of the continuity thesis and a good guide to early
developments in the Scientific Revolution.

[43] Michael J. Crowe, *Theories of the World from Antiquity to the Copernican*
Revolution (New York: Dover, 1990). Good, brief and simple account
of some of the technicalities of Ptolemaic and Copernican astronomy.

[44] Andrew Cunningham, 'Fabricius and the "Aristotle project" in Anatomical Teaching and Research at Padua', in A. Wear, R. K. French
and I. M. Lonie (eds), *The Medical Renaissance of the Sixteenth Century*
(Cambridge: Cambridge University Press, 1985), pp. 195–222. Revealing article on Harvey's Aristotelian background.

[45] Andrew Cunningham, 'William Harvey: The Discovery of the Circulation of the Blood', in Roy Porter (ed.), *Man Masters Nature: 25 Centu-*
ries of Science (London: BBC Books, 1987), pp. 65–76. A gem of an
article which clearly and succinctly explains Harvey's achievement.

[46] Peter Dear, '*Totius in verba*: Rhetoric and Authority in the Early Royal
Society', *Isis,* 76 (1985), 145–61. Clear and succinct account of the
importance of methodological claims to the success of the experimental philosophy and the Royal Society.

[47] Peter Dear, *Mersenne and the Learning of the Schools* (Ithaca, NY: Cornell
University Press, 1988). Study of a major figure in the Scientific Revolution with important lessons about changing attitudes to mathematics and experiment.

[48] Peter Dear, 'Miracles, Experiments and the Ordinary Course of Nature', *Isis*, 81 (1990), 663–83. Suggestive account of differences between English and Continental experimental method.

[49] Peter Dear, 'The Church and the New Philosophy', in S. Pumfrey, P. L. Rossi and M. Slawinski (eds), *Science, Culture and Popular Belief in Renaissance Europe* (Manchester: Manchester University Press, 1991), pp. 119–39. Excellent brief survey, includes case-study of Galileo affair.

[50] Peter Dear, *Discipline and Experience: The Mathematical Way in the Scientific Revolution* (Chicago: University of Chicago Press, 1995). Fuller restatement of material and themes in [47 and 48]. Important account of changes in intellectual authority.

[51] A. G. Debus, *The English Paracelsians* (London: Oldbourne, 1965). A history of the ideas of Paracelsus in England up to 1640.

[52] A. G. Debus, *Science and Education in the Seventeenth Century: The Webster–Ward Debate* (London: Macdonald, 1970). Introduces and reprints polemic between John Webster and Seth Ward about the curriculum in the English universities.

[53] A. G. Debus, *Man and Nature in the Renaissance* (Cambridge: Cambridge University Press, 1978). A survey of the Scientific Revolution which pays less attention to the physical sciences and more to magical and chemical ideas than do [21, 55, 235].

[54] A. G. Debus, *The French Paracelsians: The Chemical Challenge to Medical and Scientific Tradition in Early Modern France* (Cambridge: Cambridge University Press, 1991). Mostly concerned with use of chemistry in medicine but has interesting things to say about the beginnings of chemistry as a discipline in its own right.

[55] E. J. Dijksterhuis, *The Mechanization of the World Picture* (Oxford: Oxford University Press, 1961). Classic account of the Scientific Revolution, concentrating on the physical and mathematical sciences.

[56] Betty Jo Teeter Dobbs, *The Foundations of Newton's Alchemy: Or, 'The Hunting of the Greene Lyon'* (Cambridge: Cambridge University Press, 1975). Excellent study of Newton's alchemy and its background.

[57] Betty Jo Teeter Dobbs, *The Janus Faces of Genius: The Role of Alchemy in Newton's Thought* (Cambridge: Cambridge University Press, 1991). Latest re-statement of role of alchemy in Newton's thought. More advanced than [56]; superb on religious dimension.

[58] Stillman Drake, *Galileo Studies: Personality, Tradition, and Revolution* (Ann Arbor: University of Michigan Press, 1970). A collection of studies by one of the leading experts on Galileo's science.

[59] Stillman Drake, *Galileo at Work: His Scientific Biography* (Chicago: University of Chicago Press, 1978). Most detailed survey of Galileo's work.

[60] J. L. E. Dreyer, *History of Astronomy from Thales to Kepler*, 2nd edn (New York: Dover, 1953). A classic study, still useful for sorting out technical details of the astronomy.

[61] William Eamon, 'Technology as Magic in the Late Middle Ages and

the Renaissance', *Janus* (1983), 171–212. Revealing study of the role of technology in the history of natural magic.

[62] William Eamon, *Science and the Secrets of Nature: Books of Secrets in Medieval and Early Modern Culture* (Princeton: Princeton University Press, 1994). Superb study of the role of magical traditions in the formation of modern science.

[63] Norma E. Emerton, *The Scientific Reinterpretation of Form* (Ithaca, NY: Cornell University Press, 1984). A detailed account of technical developments in matter theory from the Middle Ages to the nineteenth century, but very strong on the Scientific Revolution.

[64] R. J. W. Evans, *Rudolf II and his World: A Study in Intellectual History, 1576–1612* (Oxford: Clarendon Press, 1973). Includes an excellent survey of the importance of magical traditions to Rudolf II (Ch. 6).

[65] R. J. W. Evans, *The Making of the Habsburg Monarchy, 1550–1700* (Oxford: Clarendon Press, 1979). Considers the importance of occult arts in the Habsburg Court up to the end of the seventeenth century.

[66] Annibale Fantoli, *Galileo: For Copernicus and for the Church*, trans. G. V. Coyne (Vatican City State: Vatican Observatory Publications/Notre Dame: University of Notre Dame Press, 1994). Fullest account of Galileo's difficulties with his Church since Santillana's [185].

[67] Benjamin Farrington, *Francis Bacon, Philosopher of Industrial Science* (New York: Collier Books, 1961). Classic account of Bacon's attempt to reform natural philosophy. Very useful.

[68] Mordechai Feingold, 'The Occult Tradition in the English Universities of the Renaissance: A Reassessment', in [222], pp. 73–94. A brief but suggestive survey.

[69] Mordechai Feingold, *The Mathematician's Apprenticeship: Science, Universities and Society in England, 1560–1640* (Cambridge: Cambridge University Press, 1984). Rich survey of natural philosophy in early modern Oxford and Cambridge.

[70] Lewis Feuer, *The Scientific Intellectual: The Psychological and Sociological Origins of Modern Science* (New York: Basic Books, 1963). Perhaps unfairly neglected alternative to the Puritanism-and-Science thesis.

[71] J. V. Field, *Kepler's Geometrical Cosmology* (London: Athlone, 1988). The best treatment of Kepler's cosmology, but doesn't discuss the *Astronomia nova.*

[72] Paula Findlen, *Possessing Nature: Museums, Collecting and Scientific Culture in Early Modern Italy* (Berkeley: University of California Press, 1994). A history of natural history and associated methodological developments. Also strong on the social background.

[73] James E. Force, *William Whiston: Honest Newtonian* (Cambridge: Cambridge University Press, 1985). Excellent account of a leading Newtonian.

[74] Daniel Fouke, 'Mechanical and "Organical" Models in Seventeenth-Century Explanations of Biological Reproduction', *Science in Context*, 3 (1989), 365–82. Useful survey of a problematic area for the mechanical philosophy.

[75]   Robert G. Frank, *Harvey and the Oxford Physiologists: Scientific Ideas and Social Interaction* (Berkeley: University of California Press, 1980). Superb study of early scientific collaboration and rivalry and a full account of developments in physiology.

[76]   Roger French, 'The Anatomical Tradition', in W. F. Bynum and Roy Porter (eds), *Companion Encyclopaedia of the History of Medicine*, 2 vols (London: Routledge, 1993), I, pp. 81–101. Brief survey of anatomy teaching.

[77]   Roger French, *William Harvey's Natural Philosophy* (Cambridge: Cambridge University Press, 1995). Detailed study of the development of Harvey's method and the contemporary impact of his work.

[78]   Alan Gabbey, 'Force and Inertia in the Seventeenth Century: Descartes and Newton', in [87], pp. 230–320. Useful, compressed discussion of Descartes' and Newton's views on force and inertia, but on Descartes, see also [106].

[79]   Alan Gabbey, 'Newton and Natural Philosophy', in R. C. Olby, G. N. Cantor, J. R. R. Christie and M. J. S. Hodge (eds), *Companion to the History of Modern Science* (London: Routledge, 1990), pp. 243–63. Excellent brief overview of Newton and his significance.

[80]   Alan Gabbey, 'Newton's *Mathematical Principles of Natural Philosophy*: A Treatise of Mechanics?', in P. Harman and A. Shapiro (eds), *The Investigation of Difficult things: Essays on Newton and the History of the Exact Sciences* (Cambridge: Cambridge University Press, 1992), pp. 305–22. Discusses the changing meaning of mechanics in the Scientific Revolution.

[81]   Alan Gabbey, 'Between *ars* and *philosophia naturalis*: Reflections on the Historiography of Early Modern Mechanics', in J. V. Field and F. A. J. L. James (eds), *Renaissance and Revolution: Humanists, Scholars, Craftsmen and Natural Philosophers in Early Modern Europe* (Cambridge: Cambridge University Press, 1993), pp. 133–45. Important brief discussion of the changing meaning of mechanics in the Scientific Revolution.

[82]   Galileo Galilei, *Siderius Nuncius, or the Sidereal Messenger*, trans. and ed. Albert Van Helden (Chicago: University of Chicago Press, 1989). Best edition of a seminal text.

[83]   Daniel Garber, 'Descartes' Physics', in John Cottingham (ed.), *The Cambridge Companion to Descartes* (Cambridge: Cambridge University Press, 1992), pp. 286–334. Very useful brief digest of Descartes' natural philosophy.

[84]   Daniel Garber, 'Leibniz: Physics and Philosophy', in Nicholas Jolley (ed.), *The Cambridge Companion to Leibniz* (Cambridge: Cambridge University Press, 1995), pp. 270–352. Superb brief account of Leibniz' natural philosophy.

[85]   John Gascoigne, *Cambridge in the Age of Enlightenment: Science, Religion and Politics from the Restoration to the French Revolution* (Cambridge: Cambridge University Press, 1989). Good account of the 'holy alliance'

between Newtonianism and Anglicanism, rich in detail on Cambridge.

[86] John Gascoigne, 'A Reappraisal of the Role of the Universities in the Scientific Revolution', in D. C. Lindberg and R. S. Westman (eds), *Reappraisals of the Scientific Revolution* (Cambridge: Cambridge University Press, 1990), pp. 207–60. Judicious re-appraisal of a contentious area.

[87] Stephen Gaukroger (ed.), *Descartes: Philosophy, Mathematics and Physics* (Hassocks, Sussex: Harvester Press, 1980). An anthology which concentrates on Descartes as natural philosopher.

[88] Stephen Gaukroger, 'Descartes' Project for a Mathematical Physics', in [87], pp. 97–140. Clear account of the significance of Descartes' work in geometry.

[89] Stephen Gaukroger, *Descartes: An Intellectual Biography* (Oxford: Clarendon Press, 1995). Fullest treatment of Descartes' life and work.

[90] Neal C. Gillespie, 'Natural History, Natural Theology and Social Order: John Ray and the "Newtonian Ideology"', *Journal of the History of Biology*, 20 (1987), 1–49. Detailed study of the merging of natural history and natural theology at the end of the seventeenth century and their use in social and religious apologetics.

[91] C. C. Gillispie, *The Edge of Objectivity: An Essay in the History of Scientific Ideas* (Princeton: Princeton University Press, 1960). A classic study which begins with four elegant and learned chapters on the Scientific Revolution.

[92] Penelope Gouk, 'The Harmonic Roots of Newtonian Science', in J. Fauvel, R. Flood, M. Shortland and R. Wilson (eds), *Let Newton Be!* (Oxford: Oxford University Press, 1988), pp. 101–26. Interesting account of a little-known side of Newton's work.

[93] Edward Grant, *Much Ado About Nothing: Theories of Space and Vacuum from the Middle Ages to the Scientific Revolution* (Cambridge: Cambridge University Press, 1981). A tale told by no idiot, signifying a lot more than nothing.

[94] Anita Guerrini, 'Isaac Newton, George Cheyne and the *Principia Medicinae*', in Roger French and Andrew Wear, *The Medical Revolution of the Seventeenth Century* (Cambridge: Cambridge University Press, 1989), pp. 222–45. A useful indicator of Newton's impact on medical theory.

[95] Richard W. Hadden, *On the Shoulders of Merchants: Exchange and the Mathematical Conception of Nature in Early Modern Europe* (Albany: State University of New York Press, 1994). Most recent attempt to argue for capitalist origins of 'the mathematical-mechanistic world-picture'.

[96] Roger Hahn, *The Anatomy of a Scientific Institution: The Paris Academy of Sciences, 1666–1803* (Berkeley: University of California Press, 1971). Concentrates on developments after the Scientific Revolution period, but still an important introduction to its subject.

[97] A. R. Hall, *Philosophers at War: The Quarrel between Newton and Leibniz* (Cambridge: Cambridge University Press, 1980). Fullest account of the priority dispute over calculus.

[98]   A. R. Hall, 'On Whiggism', *History of Science*, 21 (1983), 45–59. Interesting discussion of an important historiographical problem for historians of science.

[99]   A. R. Hall, *The Revolution in Science, 1500–1750* (London: Longman, 1983). The best book-length introduction to our topic.

[100]   A. R. Hall, *Isaac Newton, Adventurer in Thought* (Oxford: Blackwell, 1993). A useful biography, organised to give a good chronological account of Newton's major work, but not as full or well-rounded a portrait as Westfall's [236].

[101]   Thomas S. Hall, *History of General Physiology, 600 BC to AD 1900*, 2 vols (Chicago: University of Chicago Press, 1969). A full and useful survey of the technical issues and developments.

[102]   Thomas L. Hankins, 'Eighteenth-Century Attempts to Resolve the *vis viva* Controversy', *Isis*, 56 (1965), pp. 281–97. Good brief summary of developments.

[103]   Thomas L. Hankins, *Science and the Enlightenment* (Cambridge: Cambridge University Press, 1985). Excellent survey of science in the Enlightenment, the period immediately following the Scientific Revolution.

[104]   Thomas L. Hankins and Robert J. Silverman, *Instruments and the Imagination* (Princeton: Princeton University Press, 1995). Opening chapters of this wide-ranging survey show the links between the magical tradition and the early development of scientific instruments.

[105]   Owen Hannaway, 'Laboratory Design and the Aim of Science: Andreas Libavius versus Tycho Brahe', *Isis*, 77 (1986), 585–610. A finely nuanced study of differing conceptions of laboratories for alchemical research deriving from different beliefs about the nature of science.

[106]   Gary Hatfield, 'Force (God) in Descartes' Physics', *Studies in History and Philosophy of Science*, 10 (1979), 113–40. Important study of the nature of force in Descartes' philosophy.

[107]   Gary Hatfield, 'Metaphysics and the New Science', in D. C. Lindberg and R. S. Westman, *Reappraisals of the Scientific Revolution* (Cambridge: Cambridge University Press, 1990), pp. 93–166. Excellent but demanding account of the role of metaphysical speculation, or the lack of it, in Copernicus, Kepler, Descartes and Galileo.

[108]   Gary Hatfield, 'Descartes' Physiology and its Relation to his Psychology', in John Cottingham (ed.), *The Cambridge Companion to Descartes* (Cambridge: Cambridge University Press, 1992), pp. 335–70. Useful brief account of Descartes' ideas about the body and its relationship with mind.

[109]   John L. Heilbron, *Electricity in the Seventeenth and Eighteenth Centuries: A Study in Early Modern Physics* (Berkeley: University of California Press, 1979). Best account of its subject but also good on the physical sciences in general.

[110]   John Henry, 'Occult Qualities and the Experimental Philosophy: Active Principles in pre-Newtonian Matter Theory', *History of Science*, 24

(1986), 335–81. Suggests that Newton was not unique in his use of putative 'active principles' in matter.

[111] John Henry, 'Magic and Science in the Sixteenth and Seventeenth Centuries', in R. C. Olby et al. (eds), *Companion to the History of Modern Science* (London: Routledge, 1990), pp. 583–96. Brief and useful, if perhaps over-simplified, survey of the role of magic in the origins of modern science.

[112] John Henry, 'The Scientific Revolution in England', in R. Porter and M. Teich (eds), *The Scientific Revolution in National Context* (Cambridge: Cambridge University Press, 1992), pp. 178–210. Seeks to clarify role of religion in the shaping of the experimental philosophy in England.

[113] Christopher Hill, 'William Harvey and the Idea of Monarchy', in [227], pp. 160–81. Makes boldly suggestive claims about the impact of political developments on Harvey's way of seeing the heart and blood.

[114] William L. Hine, 'Marin Mersenne: Renaissance Naturalism and Renaissance Magic', in [222], pp. 165–76. Seeks to distinguish between naturalistic (even though occult) explanations and supposedly more demonic magical explanations. Should be compared with Clarke's essays [25; 26].

[115] R. Hooykaas, *Religion and the Rise of Modern Science* (Edinburgh: Scottish Academic Press, 1973). Classic study of the relationship between Protestant theology and early modern science.

[116] Michael Hunter, *Science and Society in Restoration England* (Cambridge: Cambridge University Press, 1981). Excellent introduction to the major themes in the historiography of late seventeenth-century English science. Includes a very useful bibliographical essay.

[117] Michael Hunter, *The Royal Society and Its Fellows, 1660–1700: The Morphology of an Early Scientific Institution*, 2nd edn (Faringdon, Oxon: British Society for the History of Science, 1994). Meticulously detailed prosopographical survey of Fellows of the Royal Society.

[118] Michael Hunter, *Establishing the New Science: The Experience of the Early Royal Society* (Woodbridge, Suffolk: Boydell Press, 1989). Important advanced collection of essays on various aspects of the early Royal Society.

[119] Michael Hunter, 'Science and Heterodoxy: An Early Modern Problem Reconsidered', in D. C. Lindberg and R. S. Westman, *Reappraisals of the Scientific Revolution* (Cambridge: Cambridge University Press, 1990), pp. 437–60. Excellent brief account of fears of atheism in seventeenth-century England.

[120] Michael Hunter (ed.), *Robert Boyle Reconsidered* (Cambridge: Cambridge University Press, 1994). Useful collection of essays on different aspects of Boyle's work and significance.

[121] Keith Hutchison, 'What Happened to Occult Qualities in the Scientific Revolution?', *Isis*, 73 (1982), 233–53. Revisionist piece on the nature of occult qualities. Essential reading for understanding the mechanical philosophy and the role of magic in the Scientific Revolution.

107

[122] Keith Hutchison, 'Towards a Political Iconology of the Copernican Revolution', in Patrick Curry (ed.), *Astrology, Science and Society* (Woodbridge, Suffolk: Boydell Press, 1987), pp. 95–141. Interesting attempt to indicate force of political symbolism in acceptance of Copernican theory.

[123] Carolyn Iltis, 'The Leibnizian-Newtonian Debates: Natural Philosophy and Social Psychology', *British Journal for the History of Science*, 6 (1973), 343–77. Fascinating account of the *vis viva* controversy showing that it was based upon a fundamental clash of world-views.

[124] James R. Jacob and Margaret C. Jacob, 'The Anglican Origins of Modern Science: The Metaphysical Foundations of the Whig Constitution', *Isis*, 71 (1980), 251–67. A suggestive essay offered as a refinement of the Puritanism-and-Science thesis.

[125] Margaret C. Jacob, *The Newtonians and the English Revolution, 1689–1720* (Ithaca, NY: Cornell University Press/Hassocks: Harvester Press, 1976). Discusses social and political context of the rise of Newtonianism. Best account of the Boyle Lectures.

[126] Nicholas Jardine, *The Birth of History and Philosophy of Science: Kepler's A Defence of Tycho against Ursus with Essays on its Provenance and Significance* (Cambridge: Cambridge University Press, 1984). Important contribution to our understanding of changing attitudes to the intellectual authority of mathematics.

[127] Nicholas Jardine, 'Epistemology of the Sciences', in C. B. Schmitt and Q. Skinner (eds), *The Cambridge History of Renaissance Philosophy* (Cambridge: Cambridge University Press, 1988), pp. 685–711. Very advanced and demanding but still convenient survey of notions of intellectual authority.

[128] Thomas Harmon Jobe, 'The Devil in Restoration Science: The Glanvill-Webster Witchcraft Debate', *Isis*, 72 (1981), 343–56. Fascinating case-study showing how experimentalism and the mechanical philosophy could be used to support belief in demonology.

[129] Nicholas Jolley, 'The Reception of Descartes' Philosophy', in John Cottingham (ed.), *The Cambridge Companion to Descartes* (Cambridge: Cambridge University Press, 1992), pp. 393–423. Brief and useful survey.

[130] Lynn Sumida Joy, *Gassendi the Atomist: Advocate of History in an Age of Science* (Cambridge: Cambridge University Press, 1987). Concentrates on Gassendi's historical scholarship to rehabilitate Epicurus' philosophy.

[131] A. G. Keller, 'Mathematicians, Mechanics and Experimental Machines in Northern Italy in the Sixteenth Century', in M. P. Crosland (ed.), *The Emergence of Science in Western Europe* (London: Macmillan, 1975), pp. 15–34. Indicates the new openings for mathematicians in the late Renaissance.

[132] Evelyn Fox Keller, *Reflections on Gender and Science* (New Haven and London: Yale University Press, 1985). Chapters 2 and 3 argue for a gendered ideology underlying the Scientific Revolution.

108

[133] Eugene M. Klaaren, *Religious Origins of Modern Science: Belief in Creation in Seventeenth-Century Thought* (Grand Rapids, Michigan: William B. Eerdmans Publishing, 1977). Particularly good on Robert Boyle and J. B. van Helmont.

[134] Alexandre Koyré, *The Astronomical Revolution: Copernicus, Kepler, Borelli*, trans. R. E. W. Maddison (London: Methuen, 1973). A detailed technical account but with lots of fascinating material, written by a founding father of the history of science.

[135] Thomas S. Kuhn, *The Copernican Revolution: Planetary Astronomy in the Development of Western Thought* (Cambridge, Mass.: Harvard University Press, 1957). Classic study of Copernicus' innovations.

[136] Thomas S. Kuhn, 'Mathematical versus Experimental Traditions in the Development of Physical Science', in idem, *The Essential Tension: Selected Studies in Scientific Tradition and Change* (Chicago: University of Chicago Press, 1977), pp. 31–65. A suggestive essay proposing that the Scientific Revolution is best understood by dividing 'classical physical sciences' (amenable to mathematical analysis) from newly emergent Baconian (empirical) sciences.

[137] W. R. Laird, 'Patronage of Mechanics and Theories of Impact in Sixteenth-Century Italy', in [155], pp. 51–66. Shows how the interests of patrons led to the introduction of theories of impact, important for the mechanical philosophy, into traditional mechanics.

[138] James M. Lattis, *Between Copernicus and Galileo: Christoph Clavius and the collapse of Ptolemaic Cosmology* (Chicago: University of Chicago Press, 1994). Major study of a leading Jesuit natural philosopher.

[139] David Lindberg, *The Beginnings of Western Science: The European Scientific Tradition in Philosophical, Religious, and Institutional Context, 600 BC to AD 1450* (Chicago: University of Chicago Press, 1992). Excellent introduction to ancient and medieval science.

[140] Arthur O. Lovejoy, *The Great Chain of Being: A Study of the History of an Idea* (New York: Harper & Row, 1960). Classic study of intellectualist theology.

[141] David S. Lux, *Patronage and Royal Science in Seventeenth-Century France: The Académie de Physique in Caen* (Ithaca, NY: Cornell University Press, 1989). Enlightening account of a less well-known early scientific institution.

[142] James E. McClellan III, *Science Reorganized: Scientific Societies in the Eighteenth Century* (New York: Columbia University Press, 1985). Best and fullest survey of scientific societies in the Scientific Revolution and after.

[143] Charles J. McCracken, *Malebranche and British Philosophy* (Oxford: Clarendon Press, 1983). Best book on Malebranche and his influence.

[144] J. E. McGuire and P. M. Rattansi, 'Newton and the "Pipes of Pan"', *Notes and Records of the Royal Society of London*, 21 (1966), 108–43. A fascinating and important study of Newton's belief in the ancient wisdom of Neoplatonic and Pythagorean traditions.

[145] Julian Martin, *Francis Bacon, the State, and the Reform of Natural Philosophy* (Cambridge: Cambridge University Press, 1992). Best account of Bacon's philosophical reforms. Argues that his statesmanship is crucial to understanding his aims and methods.

[146] Otto Mayr, *Authority, Libery and Automatic Machinery in Early Modern Europe* (Baltimore and London: Johns Hopkins University Press, 1986). Superb account of ideological influences upon the development of technology.

[147] Christoph Meinel, 'Early Seventeenth-Century Atomism: Theory, Epistemology, and the Insufficiency of Experiment', *Isis*, 79 (1988), 68–103. Useful account of early developments in the revival of atomism.

[148] Everett Mendelsohn, *Heat and Life: The Development of the Theory of Animal Heat* (Cambridge, Mass.: Harvard University Press, 1964). Good survey of a major aspect of the life sciences.

[149] J. Andrew Mendelsohn, 'Alchemy and Politics in England, 1649–1665', *Past and Present*, 135 (1992), 30–78. A survey of the political and rhetorical uses of alchemy which cautions against supposedly fixed affiliations between alchemy and radical religious and political views.

[150] Carolyn Merchant, *The Death of Nature: Women, Ecology and the Scientific Revolution* (San Francisco: Harper & Row, 1980). Best feminist analysis of the Scientific Revolution. Provides a suggestive analysis of the social background to the new philosophies.

[151] Robert K. Merton, *Science, Technology and Society in Seventeenth-Century England* (New York: Howard Fertig, 1970). Originally published in 1938, this is the Merton Thesis (important version of the Puritanism-and-Science thesis).

[152] W. E. K. Middleton, *The Experimenters: A study of the Accademia del Cimento* (Baltimore: Johns Hopkins University Press, 1971). A study of one of the very earliest scientific societies.

[153] Ron Millen, 'The Manifestation of Occult Qualities in the Scientific Revolution', in M. J. Osler and P. L. Farber (eds), *Religion, Science and Worldview: Essays in Honor of Richard S. Westfall* (Cambridge: Cambridge University Press, 1985), pp. 185–216. A useful pendant to [121], with some fascinating additional material.

[154] Bruce T. Moran, *The Alchemical World of the German Court: Occult Philosophy and Chemical Medicine in the Circle of Moritz of Hessen* (Stuttgart: Franz Steiner Verlag, 1991). Detailed case-study of the role of magical world-views in European courts.

[155] Bruce T. Moran (ed.), *Patronage and Institutions: Science, Technology and Medicine at the European Court, 1500–1750* (Woodbridge, Suffolk: Boydell Press, 1991). Anthology of uniformly excellent papers on issues of science and patronage.

[156] Bruce T. Moran, 'Patronage and Institutions: Courts, Universities, and Academies in Germany; an Overview 1550–1750', in [155], pp. 169–84. A useful survey of the range of opportunities for natural philosophers.

[157]   Lotte Mulligan, 'Civil War Politics, Religion and the Royal Society', in [227], pp. 317–39. A critique of the Puritanism-and-Science thesis which points to evidence for interest in science displayed by royalist Anglican gentlemen.

[158]   Lotte Mulligan, 'Puritans and English Science: A Critique of Webster', *Isis*, 71 (1980), 457–69. A sequel to [157] which argues that supposedly Puritanical traits were in fact common among non-Puritans.

[159]   William R. Newman, 'Boyle's Debt to Corpuscular Alchemy', in [120], pp. 107–18. Succinct account of a little-known alchemical tradition and Boyle's use of it.

[160]   Francis Oakley, *Omnipotence, Covenant and Order: An Essay in the History of Ideas from Abelard to Leibniz* (Ithaca, NY: Cornell University Press, 1984). Complements [140], discusses Boyle and Leibniz.

[161]   Richard Olson, *Science Deified and Science Defied: The Historical Significance of Science in Western Culture*, Vol. 2 (Berkeley: University of California Press, 1991). Useful survey of some of the ways in which science affected the wider culture.

[162]   C. D. O'Malley, *Andreas Vesalius of Brussels, 1514–1564* (Berkeley: University of California Press, 1965). Still the only book-length treatment of Vesalius.

[163]   Martha Ornstein, *The Role of Scientific Societies in the Seventeenth Century*, 3rd edn (Chicago: University of Chicago Press, 1938). Outdated but still useful introduction to the most prominent scientific societies.

[164]   Margaret J. Osler, *Divine Will and the Mechanical Philosophy: Gassendi and Descartes on Necessity and Contingency in the Created World* (Cambridge: Cambridge University Press, 1994). Shows the importance of theological views in the development of the methodologies of early modern science.

[165]   Walter Pagel, *William Harvey's Biological Ideas* (Basel: 1967). Classic account of Harvey's Aristotelianism and other aspects of the background to his work.

[166]   Walter Pagel, *Joan Baptista van Helmont: Reformer of Science and Medicine* (Cambridge: Cambridge University Press, 1982). Fullest survey of van Helmont and his work.

[167]   L. C. Palm and H. A. M. Snelders (eds), *Antoni van Leeuwenhoek, 1632–1723* (Amsterdam: Rodopi, 1982). Collection of articles on one of the leading microscopists.

[168]   Olaf Pedersen, *Early Physics and Astronomy: A Historical Introduction*, 2nd edn (Cambridge: Cambridge University Press, 1993). A clearly simplified guide to technical aspects of the astronomical revolution and its background.

[169]   Richard H. Popkin, 'Newton's Biblical Theology and his Theological Physics', in P. B. Scheurer and B. Debrock (eds), *Newton's Scientific and Philosophical Legacy* (Dordrecht: Kluwer Academic, 1988), pp. 81–97. Emphasizes religious significance and use of Newton's natural philosophy.

111

[170]   Roy Porter, 'The Scientific Revolution: A Spoke in the Wheel?', in R. Porter and M. Teich (eds), *Revolution in History* (Cambridge: Cambridge University Press, 1986), pp. 290–316. Succinct discussion of the historians' notion of the Scientific Revolution.

[171]   Lawrence Principe, 'Boyle's Alchemical Pursuits', in [120], pp. 91–105. Useful brief survey.

[172]   Andrew Pyle, 'Animal Generation and the Mechanical Philosophy: Some Light on the Role of Biology in the Scientific Revolution', *History and Philosophy of the Life Sciences*, 9 (1987), 225–54. Argues for the importance of work in the biomedical sciences for the establishment of the mechanical philosophy.

[173]   John H. Randall, 'The Development of the Scientific Method in the School of Padua', *Journal of the History of Ideas*, 1 (1940), 177–206. A study of the flourishing of Aristotelianism at Padua which supports the continuity thesis.

[174]   P. M. Rattansi, 'Paracelsus and the Puritan Revolution', *Ambix*, 11 (1963), 24–32. Succinct account of affiliations between political radicalism and Paracelsianism in Interregnum England.

[175]   C. E. Raven, *John Ray, Naturalist: His Life and Works* (Cambridge: Cambridge University Press, 1942). Classic study of a leading naturalist.

[176]   Graham Rees, 'Francis Bacon's Semi-Paracelsian Cosmology', *Ambix*, 22 (1975), 81–101. Important reconstruction of Bacon's idiosyncratic natural philosophy.

[177]   Vasco Ronchi, *The Nature of Light: An Historical Survey* (London: Heinemann, 1970). History of theories of light and optics from the Ancient Greeks to the nineteenth century but with a major focus on developments in the Scientific Revolution.

[178]   Hugh Trevor Roper, 'The Paracelsian Movement', in idem, *Renaissance Essays* (London: Secker & Warburg, 1985; Fontana, 1986), pp. 149–99. Best European-wide survey of the socio-political background to the spread of Paracelsianism.

[179]   G. MacDonald Ross, *Leibniz* (Past Masters, Oxford: Oxford University Press, 1984). Lightning survey of major aspects of Leibniz's work. Perhaps too compressed.

[180]   Paolo Rossi, *Francis Bacon: From Magic to Science* (London: Routledge & Kegan Paul, 1968). Details the likely sources for Bacon's ideas on the reform of knowledge.

[181]   M. J. S. Rudwick, *The Meaning of Fossils: Episodes in the History of Palaeontology* (Chicago and London: University of Chicago Press, 1972). The first two chapters have much to say on the history of natural history in our period.

[182]   Edward G. Ruestow, 'Piety and the Defense of the Natural Order: Swammerdam on Generation', in M. J. Osler and P. L. Farber (eds), *Religion, Science and Worldview: Essays in Honor of Richard S. Westfall* (Cambridge: Cambridge University Press, 1985), pp. 217–39. A case-

study of the links between science and religion and a useful study of an important natural philosopher.

[183] A. I. Sabra, *Theories of Light from Descartes to Newton* (Cambridge: Cambridge University Press, 1981). A major account of technical developments in optics.

[184] Danton B. Sailor, 'Moses and Atomism', *Journal of the History of Ideas*, 25 (1964), 3–16. Fascinating account of efforts to make atomism more religiously respectable.

[185] Georgio de Santillana, *The Crime of Galileo* (Chicago: University of Chicago Press, 1955). Rich and full account of the 'Galileo affair'. Emphasizes conspiratorial elements in the story; compare with Fantoli [66].

[186] Lisa T. Sarasohn, 'A Science Turned Upside Down: Feminism and the Natural Philosophy of Margaret Cavendish', *Huntington Library Quarterly*, 47 (1984), 289–307. Good brief account of Margaret Cavendish's work and its feminist intentions.

[187] Rose-Mary Sargent, *The Diffident Naturalist: Robert Boyle and the Philosophy of Experiment* (Chicago: University of Chicago Press, 1995). Excellent book-length treatment of Boyle's methodology.

[188] Simon Schaffer, 'Godly Men and Mechanical Philosophers: Souls and Spirits in Restoration Natural Philosophy', *Science in Context*, 1 (1987), 55–86. Argues that the mechanical philosophy was used to establish and define the realm of immaterial souls and spirits.

[189] Simon Schaffer, 'Newtonianism', in R. C. Olby, G. N. Cantor, J. R. R. Christie and M. J. S. Hodge (eds), *Companion to the History of Modern Science* (London: Routledge, 1990), pp. 610–26. Useful pointer to extent of Newton's influence in eighteenth and nineteenth centuries.

[190] Londa Schiebinger, *The Mind has No Sex? Women in the Origins of Modern Science* (Cambridge, Mass.: Harvard University Press, 1989). A study of the exclusion of women from science in the early modern period.

[191] Charles B. Schmitt, 'Science in the Italian Universities in the Sixteenth and Early Seventeenth Centuries', in M. Crosland (ed.), *The Emergence of Science in Western Europe* (London: Macmillan, 1975), pp. 35–56. Uses Italian material to caution against neglecting the role of the universities in the Scientific Revolution.

[192] Charles B. Schmitt, *Aristotle in the Renaissance* (Cambridge, Mass.: Harvard University Press, 1983). Important survey of the continued flourishing of different versions of Aristotelianism into the early modern period.

[193] John A. Schuster, 'The Scientific Revolution', in R. C. Olby et al. (eds), *Companion to the History of Modern Science* (London: Routledge, 1990), pp. 217–42. An excellent short account of our theme, both historically and historiographically sophisticated.

[194] Michael Segre, *In the Wake of Galileo* (New Brunswick: Rutgers University Press, 1991). A study of Galileo's immediate followers in Italy.

[195] Steven Shapin, 'Social Uses of Science', in G. Rousseau and R. Porter (eds), *The Ferment of Knowledge* (Cambridge: Cambridge University Press,

1980), pp. 93–139. A manifesto for the application of sociology of knowledge to historiography.

[196]   Steven Shapin, 'Of Gods and Kings: Natural Philosophy and Politics in the Leibniz-Clarke Disputes', *Isis*, 72 (1981), 187–215. Argues for the need to recognize the role of politics in the intellectual rivalry between Newton and Leibniz.

[197]   Steven Shapin, 'The House of Experiment in Seventeenth-Century England', *Isis*, 79 (1988), 373–404. Brief account of some of the themes in [200]. Shows how the site and situation in which experiments were performed was used to defend the legitimacy of the experimental method.

[198]   Steven Shapin, 'Who was Robert Hooke?', in M. Hunter and S. Schaffer (eds), *Robert Hooke: New Studies* (Woodbridge, Suffolk: Boydell Press, 1989), pp. 253–85. Demonstrates that who you were was relevant to how your scientific work was regarded.

[199]   Steven Shapin, 'Discipline and Bounding: The History and Sociology of Science as Seen Through the Externalism-Internalism Debate', *History of Science*, 30 (1992), 333–69. A history of the externalism-internalism debate in historiography of science, and a forceful plea for a historicist sociology of knowledge approach.

[200]   Steven Shapin, *A Social History of Truth: Civility and Science in Seventeenth Century England* (Chicago: Chicago University Press, 1994). Fascinating account of the role of trust in scientific knowledge.

[201]   Steven Shapin and Simon Schaffer, *Leviathan and the Air-Pump: Hobbes, Boyle and the Experimental Life* (Princeton: Princeton University Press, 1985). Important study of the dispute between Hobbes and Boyle with interesting things to say about the nature of science, the experimental method, and discipline boundaries.

[202]   Barbara J. Shapiro, 'Latitudinariansim and Science in Seventeenth-Century England', in [227], pp. 286–316. Influential statement of the major alternative to the Puritanism-and-Science thesis.

[203]   Barbara J. Shapiro, *Probability and Certainty in Seventeenth-Century England: A Study of the Relationships between Natural Science, Religion, History, Law and Literature* (Princeton: Princeton University Press, 1983). A rich and ambitious survey of its themes.

[204]   Michael Sharratt, *Galileo, Decisive Innovator* (Oxford: Blackwell, 1994). A judiciously simplified account which covers all major aspects of Galileo's life and work.

[205]   William R. Shea, *Galileo's Intellectual Revolution* (London: Macmillan, 1972). Good brief account of the development of Galileo's scientific method.

[206]   William R. Shea, 'Galileo and the Church', in D. C. Lindberg and R. Numbers (eds), *God and Nature: Historical Essays on the Encounter between Christianity and Science* (Berkeley: University of California Press, 1986), pp. 114–35. Succinct account of a complex affair.

[207] William R. Shea, *The Magic of Numbers and Motion: The Scientific Career of René Descartes* (Canton, Mass.: Science History Publications, 1991). Extremely useful survey of Decartes' contribution to the history of science, but perhaps misses the metaphysical dimension of his work.

[208] Phillip R. Sloan, 'Natural History, 1670–1802', in R. C. Olby, G. N. Cantor, J. R. R. Christie and M. J. S. Hodge (eds), *Companion to the History of Modern Science* (London: Routledge, 1990), pp. 295–313. In spite of its title this includes succinct introductory material on the Renaissance.

[209] Bruce Stephenson, *Kepler's Physical Astronomy* (Princeton: Princeton University Press, 1994). Most recent, and most forceful statement of the role of physical speculation in Kepler's astronomy.

[210] Bruce Stephenson, *The Music of the Heavens: Kepler's Harmonic Astronomy* (Princeton: Princeton University Press, 1994). A study of theories of celestial harmony and Kepler's use of them.

[211] Larry Stewart, *The Rise of Public Science: Rhetoric, Technology, and Natural Philosophy in Newtonian Britain, 1660–1750* (Cambridge: Cambridge University Press, 1992). Richly detailed examination of religious and commercial aspects of the rise of Newtonianism.

[212] Alice Stroup, *A Company of Scientists: Botany, Patronage, and Community at the Seventeenth-Century Parisian Royal Academy of Sciences* (Berkeley: University of California Press, 1990). A useful study of the French Academy of Sciences, and of botany in seventeenth-century science.

[213] Mary Terrall, 'Emilie du Châtelet and the Gendering of Science', *History of Science*, 33 (1995), 283–310. Revealing account of the difficulties facing women in science at this period.

[214] Jim Tester, *A History of Western Astrology* (Woodbridge, Suffolk: Boydell Press, 1987). Excellent historical survey of astrology and its significance.

[215] Victor E. Thoren, *Tycho Brahe: The Lord of Uraniborg* (Cambridge: Cambridge University Press, 1990). Most up-to-date intellectual biography of this crucially important figure.

[216] Lynn Thorndike, *A History of Magic and Experimental Science*, 8 vols (New York: Columbia University Press, 1923–58). Monumental survey of the links between experimentalism and magic.

[217] E. M. W. Tillyard, *The Elizabethan World Picture* (London: Chatto & Windus, 1943). Classic account of correspondences between different aspects of the cosmos in Elizabethan thought.

[218] Nicholas Tyacke, 'Science and Religion at Oxford before the Civil War', in D. H. Pennington and K. Thomas (eds), *Puritans and Revolutionaries: Essays in Seventeenth-Century History Presented to Christopher Hill* (Oxford: Clarendon Press, 1978), pp. 73–93. An attack on the Puritanism-and-Science thesis arguing for links between High Church Anglicanism and science.

[219] Albert Van Helden, 'The Birth of the Modern Scientific Instrument',

in John G. Burke (ed.), *The Uses of Science in the Age of Newton* (Berkeley: University of California Press, 1983), pp. 49–84. Useful essay on the significance of scientific instruments.

[220]   Albert Van Helden, 'Telescopes and Authority from Galileo to Cassini', *Osiris*, 9 (1994), 9–29. Pointing out that the validity and reliability of telescopic observations had to be established, this article shows the difficulties facing early advocates of the new instrument.

[221]   Andrew G. Van Melsen, *From Atomos to Atom: The History of the Concept Atom* (New York: Harper & Brothers, 1960). A useful survey of atomistic philosophies which includes a study of the early modern revival.

[222]   Brian Vickers (ed.), *Occult and Scientific Mentalities in the Renaissance* (Cambridge: Cambridge University Press, 1984). An indispensable anthology on the role of magic in the origins of modern science.

[223]   Brian Vickers, 'Analogy versus Identity: The Rejection of Occult Symbolism, 1580–1650', in [222], pp. 95–163. Shows how occultist assumptions about the use and nature of symbols were displaced by new classifications based on observation and experience.

[224]   William A. Wallace, *Galileo and his Sources: The Heritage of the Collegio Romano in Galileo's Science* (Princeton: Princeton University Press, 1984). Argues for the influence of the Jesuits at the Collegio Romano on Galileo early in his career.

[225]   Deborah Warner, 'Terrestrial Magnetism: For the Glory of God and the Benefit of Mankind', *Osiris*, 9 (1994), 67–84. Shows the interplay of natural philosophical and more pragmatic concerns in the development of terrestrial magnetism and magnetic instruments.

[226]   Andrew Wear, 'The Heart and Blood from Vesalius to Harvey', in R. C. Olby, G. N. Cantor, J. R. R. Christie and M. J. S. Hodge (eds), *Companion to the History of Modern Science* (London: Routledge, 1990), pp. 568–82. Useful brief survey.

[227]   Charles Webster (ed.), *The Intellectual Revolution of the Seventeenth Century* (London: Routledge & Kegan Paul, 1974). Includes an important set of papers for and against Puritanism-and-Science thesis.

[228]   Charles Webster, *The Great Instauration: Science, Medicine and Reform 1626–1660* (London: Duckworth, 1975). Magisterial account of science in England in the Interregnum. Major contribution to Puritanism-and-Science thesis.

[229]   Charles Webster, 'Alchemical and Paracelsian Medicine', in idem (ed.), *Health, Medicine and Mortality in the Sixteenth Century* (Cambridge: Cambridge University Press, 1979), pp. 301–34. Useful brief survey.

[230]   Charles Webster, *From Paracelsus to Newton: Magic and the Making of Modern Science* (Cambridge: Cambridge University Press, 1982). Brief but densely packed survey of the place of astrology, natural magic and witchcraft in the Scientific Revolution.

[231]   Charles Webster, 'Puritanism, Separatism, and Science', in D. C. Lindberg and R. Numbers (eds), *God and Nature: Historical Essays on the Encoun-*

ter between Christianity and Science (Berkeley: University of California Press, 1986), pp. 192–217. Brief re-statement of views expressed in Conclusion of [228].

[232] Charles Webster, 'Paracelsus: Medicine as Popular Protest', in O. P. Grell and A. Cunningham (eds), *Medicine and the Reformation* (London and New York: Routledge, 1993), pp. 57–77. Argues for the close links between Paracelsus' medical theories and his reformist religious views.

[233] R. S. Westfall, *Force in Newton's Physics: The Science of Dynamics in the Seventeenth Century* (History of Science Library, London: Macdonald, 1971). The subtitle gives a better idea of the book's coverage. Convenient one-volume survey of a crucially important but complex issue.

[234] R. S. Westfall, *Science and Religion in Seventeenth-Century England* (Ann Arbor: University of Michigan Press, 1973). Full survey of natural theology in England.

[235] R. S. Westfall, *The Construction of Modern Science: Mechanisms and Mechanics* (Cambridge: Cambridge University Press, 1977). A survey of the Scientific Revolution concentrating on the development of the mechanical philosophy.

[236]. R. S. Westfall, *Never at Rest: A Biography of Isaac Newton* (Cambridge: Cambridge University Press, 1980). A superb biography, combined with an excellent account of Newton's work.

[237] R. S. Westfall, 'Newton and Alchemy', in [222], pp. 315–35. Succinct statement of claim that Newton's concept of force derived from alchemy.

[238] Robert S. Westman, 'The Astronomer's Role in the Sixteenth Century: A Preliminary Survey', *History of Science*, 18 (1980), 105–47. Important study of sixteenth-century debates about the intellectual authority of astronomers compared to natural philosophers. Has interesting things to say about discipline boundaries.

[239] Robert S. Westman, 'The Copernicans and the Churches', in D. C. Lindberg and R. Numbers (eds), *God and Nature: Historical Essays on the Encounter between Christianity and Science* (Berkeley: University of California Press, 1986), pp. 76–113. Excellent brief survey of religious reactions to Copernicus.

[240] G. Whitteridge, *William Harvey and the Circulation of the Blood* (London: Macdonald, 1971). A detailed study of the background to, and development of, Harvey's theory of the circulation. Should be read in connection with [165].

[241] Catherine Wilson, *The Invisible World: Early Modern Philosophy and the Invention of the Microscope* (Princeton: Princeton University Press, 1995). Excellent survey of early microscopy and its impact on natural philosophy.

[242] David E. Wolfe, 'Sydenham and Locke on the Limits of Anatomy', *Bulletin of the History of Medicine*, 35 (1961), 193–200. Illustrates and explains resistance to the use of the microscope in natural philosophy.

117

[243]   Paul B. Wood, 'Methodology and Apologetics: Thomas Sprat's *History of the Royal Society*', *British Journal for the History of Science*, 13 (1980), 1–26. A study of the significance of the religious context for understanding the professed methodology of the Royal Society.

[244]   Joella G. Yoder, *Unrolling Time: Christiaan Huygens and the Mathematization of Nature* (Cambridge: Cambridge University Press, 1988). A major study of an important figure and an important case-study of the development of mathematical physics.

[245]   Edgar Zilsel, 'The Origins of William Gilbert's Scientific Method', *Journal of the History of Ideas*, 2 (1941), 1–32. Early but still influential essay on the origins of the experimental method.

Glossary

This glossary gives brief explanations of the technical scientific terms, the major historical references and the significant historiographical concepts. The scientific and historiographical entries include references to the bibliography for those wishing to pursue the matter further. For further brief discussions of the scientific terms see the *Dictionary of the History of Science*, ed. W. F. Bynum, E. J. Browne and Roy Porter (London: Macmillan, 1983).

active principles – principles residing in bodies which are supposed to be the cause of various activities of those bodies, such as gravitational attraction, and fermentation. A Newtonian notion, prefigured in earlier thinkers [56, 57, 110, 236].

alchemy – ancient art aimed at producing perfection (manifested by turning base metal into gold, for example) by exploiting the ways in which different substances can be made to react with one another to produce new substances. In many ways, therefore, it was a kind of proto-chemistry but because of the search for perfection it was usually overlaid with mystical significance [56, 57, 237].

animalculism – belief, deriving from Antony van Leeuwenhoek's discovery of spermatozoa (1677), that an organism's progeny are carried, preformed, in the male seed [235, 241].

animism – belief that natural objects (even those seemingly devoid of life) are endowed with souls (and therefore intelligence of some kind).

apologetics – the rational defence and justification of theology and other aspects of religion.

Aristotelian – derived from or based upon the work of Aristotle (384–322 BC), most influential of the Ancient Greek philosophers upon natural philosophy. Should not be seen as monolithic, there were numerous refinements and variations upon the basic themes [192].

astrology – the study and interpretation of the influence of the stars upon human and other earthly affairs. Widely believed and philosophically justified throughout the Middle Ages and the Renaissance, but in decline by the end of the seventeenth century [38, 214].

astronomy – the study and interpretation of the movements of the heavenly bodies. A practical art useful for calendrical determinations and in

119

navigation, astrology and cosmology; and one of the mixed mathematical sciences [43, 60, 134, 135, 138, 168, 215, 238].

atomism – ancient philosophical system which explains all physical phenomena in terms of the motions, combinations and arrangements of indivisible particles of matter, called atoms [221, 147].

Baconian – derived from or based upon the doctrines of Francis Bacon (1561–1626). Often used to designate observational, classificatory and empirical procedures [136, 145, 180].

bête-machine – Cartesian conception of an animal as a complex automaton [19, 89, 235].

Calvinism – the Protestant religious system developed by Jean Calvin (1509–64), sometimes equated with Puritanism [15, 151, 158, 227, 231].

Cartesian – derived from or based upon the work of René Descartes (1596–1650) [87, 89, 207].

centrifugal force – force outwards from the centre experienced by a rotating body [233, 244].

centripetal force – force inwards to a centre, such as the force of gravity [233, 244].

continuist – historiographical position which asserts that early modern innovations in science can be shown to have grown out of medieval natural philosophy [139].

Copernican – derived from or based upon the theory of Nicolaus Copernicus (1473–1543). Often used loosely to refer to any system in which the earth is in motion around the sun. So, Johannes Kepler (1571–1630) can be described as a Copernican even though his elliptical planetary orbits mark him out as crucially different in his astronomy [43, 60, 135].

corpuscularism – philosophical position similar to atomism but in which the fundamental particles of matter are divisible, or not proven to be indivisible [99, 235].

correspondences – supposed links between different bodies in corresponding positions on the 'Great Chain of Being'. The Sun, noblest of the heavenly bodies, was held to correspond to gold, the noblest metal. It was believed that corresponding bodies could influence one another and were used in magical procedures to bring about desired ends [217, 111].

cosmology – the study of the structure and system of the universe. Seen as a science which was, or should be, supported by the art of astronomy [43, 71, 134, 238].

Counter-Reformation/Counter-Reforming – movement/measures initiated by the Roman Catholic Church to counteract the effects of the Protestant Reformation and to win back converts [7, 66, 185, 204, 206].

deductive logic – pre-eminent form of logic for Aristotle, because of its ability to generate certain conclusions. Based upon the different kinds of syllogism [50].

deism – a belief in the principles of natural religion and natural theology. The deist accepts on supposedly naturalistic or rational grounds the existence of God, the immortality of the soul and other fundamental aspects of religious belief, but denies or ignores many of the precepts derived from the Scriptures (such as the virgin birth, the divinity, Resurrection and Ascension of Christ, the Holy Trinity) [125, 161, 234].

deferent – in astronomy the major circle used, in combination with an epicycle, to define the motion of a planet around the centre of the world system. See epicycle [43, 60, 135, 168].

discipline boundaries – intellectually conceived demarcations between different specialist subjects, such as between botany and zoology. Many aspects of the Scientific Revolution can be seen as the result of changes in the way these boundaries were drawn. Consider, for example, changes to the boundary between astronomy and cosmology [50, 126, 135, 238], between mechanics and the natural philosophy of motion [9, 47, 50, 80, 81, 137], between natural magic and natural philosophy [26, 39, 61, 62, 111, 153, 180, 216], or natural history and natural philosophy [8, 48, 72, 241].

dualism – belief that soul and body are categorically distinct entities [15, 89, 108].

dynamics – originally coined by G. W. Leibniz to refer to his own way of explaining things in terms of his concept of force. Usually used, as here, to refer to any attempt to explain physical phenomena in terms of the operation of forces [78, 80, 81, 84, 102, 233]. Compare with kinematics.

early modern period – used loosely to refer to the period immediately following the Renaissance. As used here it should be taken to begin in the sixteenth century and to cover the whole of our period.

eccentric – displaced from the centre. Supposing the earth to be somewhat off-centre was a simple technique used in astronomy to partially account for observed variations in speed and brightness (and therefore distance from the earth) of the planets [43, 60, 135, 168].

emboîtement – a preformationist theory which supposes that all subsequent generations are encapsulated within the egg (ovism) or the sperm (animalculism) of any given generation. The ovist believes, for example, that a female contains eggs, some of which contain miniscule preformed females who have eggs, some of which contain even more miniscule preformed females who have eggs, and so on [101, 182, 235, 172].

epicycle – in astronomy a lesser circle about which a planet is taken to rotate while that circle itself moves around a major circle (the deferent) centred at or near the centre of the world system. The combined rotations of planet and epicycle enabled astronomers to accommodate observed variations in speed and brightness (and therefore distance from the earth) of the planets and their retrograde motions (planets were envisaged to loop-the-loop) without deviating from the Ancient Greek stricture that the motion of the heavenly bodies must be uniform (unchanging) and perfectly circular [43, 60, 135, 168].

epigenesis – theory of generation which assumes that embryonic development takes place gradually from previously undifferentiated material. Although advocated by Harvey and Descartes, it seemed inexplicable from the perspective of the mechanical philosophy and tended to give place to rival preformationist theories [101, 235, 172].

epistemology – the study and theory of how knowledge is acquired and confirmed (by sensory experience, for example, or by the use of reason) [89, 107, 127, 201].

equant – an imaginary point in space, some distance from the centre of a planet's deferent, from which the motion of a planet (or rather the motion of the imaginary centre of its epicycle) would seem to be uniform and unchanging. An innovation of Ptolemy's (AD 90–168), the implication was that the centre of the epicycle was moving around the deferent *not* uniformly but with varying speeds. Decried as a deviation from Ancient Greek precepts and as physically unaccountable (how could the centre of an epicycle maintain uniform motion with respect to an eccentric point?), it nevertheless proved useful in accounting for observed planetary motions [43, 60, 135, 168].

experimental philosophy – name given to the natural philosophy promoted by late seventeenth-century English thinkers, particularly the leading members of the Royal Society, in which practitioners claimed to establish uncontestable matters of fact by Baconian experimentalism. Unlike Continental mechanical philosophers, the experimental philosophers could accept the existence of occult qualities, such as gravitational or magnetic attraction, or the 'spring' of the air, as experimentally established matters of fact [48, 110, 112, 201, 243].

external history – history writing which seeks to explain the formation of innovatory ideas in science and their acceptance or non-acceptance by contemporaries in terms of social, political, religious and other cultural influences. Frequently criticized, usually naïvely, for not paying sufficient attention to the internal dynamics of the science in question, its internal logic, the supposedly unassailable demonstrations of experiment, and the supposedly unambiguous dictates of nature itself. Compare with internal history [199].

force – an operator capable of bringing about a change of motion in a body. In the mechanical philosophy usually held to derive from the motion of bodies, hence usually synonymous with 'force of impact' or 'force of percussion'. Also used, chiefly outside the mechanical philosophy (but also, for example, by Newton) to refer to operations capable of acting at distances and hence regarded as 'spiritual' or 'occult' [57, 78, 83, 84, 102, 233].

Galenic – derived from or based upon the work of Galen (AD 129–199), the ancient medical systematist who proved to be as influential in medicine as Aristotle in natural philosophy [77, 226, 240].

geocentric – used to denote an astronomical system in which the earth is at, or near, the centre [43, 60, 135, 168].

geostatic – used to denote an astronomical system in which the earth is stationary [43, 60, 135, 168].

122

Glorious Revolution – so-called revolution of 1688 which saw William and Mary established as joint monarchs after the abdication of the Catholic convert, James II.

Great Chain of Being – shorthand way of referring to the traditional and widespread belief that God had arranged all created creatures in an unbroken hierarchy, so that every animal, for example, had its allotted place in the hierarchy, inferior to what was above and superior to what was below [140, 160, 217].

heliocentric – used to denote an astronomical system in which the Sun is at (or very near) the centre [43, 60, 135].

Hellenistic Greek – here used to denote a later Greek thinker, flourishing after the death of Alexander the Great (323 BC) or, more to the point, after Aristotle (*d.* 322 BC).

Hermetic Tradition – tradition of thought deriving from the Hermetic Corpus, writings supposedly written in deepest antiquity by Hermes Trismegistus but now known to have been written in the early Christian era. Has enjoyed an important place in the historiography of the Scientific Revolution but should really be seen as merely one aspect of the broader Neoplatonic tradition [39].

historiography – the writing of history. Historiographical and historical do not, therefore, mean the same thing. Galileo's historical significance derives from who he was and what he did. His historiographical significance derives from the great attention that he has been paid by historians of science. Historiographical controversies are conducted by historians, historical controversies are conducted by important people [199].

homocentric – having the same centre. Refers to the nesting of the heavenly spheres, as portrayed in the Aristotelian world picture and in certain Ancient Greek astronomical systems, in which all rotations are centred upon the Earth. Distinct from the many-centred Ptolemaic system of epicycles centred upon deferents [43, 60, 135, 168].

humoral pathology – the study of illness in terms of a disturbance to the normal balance of the four humours in the body [101, 139].

humours – four in number, the bodily equivalent of the so-called four elements (fire, air, water and earth), loosely identified with four bodily fluids: choler or yellow bile, blood, phlegm and black bile or melancholy. The identification was by no means rigid: any watery substance in the body, for example, would be seen as phlegm, or predominantly phlegm [101, 139].

hydrostatics – the study of bodies floating or suspended in fluids. One of the mixed mathematical sciences [58, 59].

impetus theory – developed in the Middle Ages by Jean Buridan (*c.* 1295–1358) and adopted by Galileo Galilei (1564–1642) as an alternative to Aristotelian accounts of motion, the theory supposed that projectiles continue to move after leaving contact with their projectors (problematic for Aristotle) as the result of an imparted *impetus*. Impetus is used up during the projectile's flight [99, 139, 224].

induction – a form of inference usually deemed to be inferior to deduction, because of its lack of certainty and its failure to provide causative explanations, but championed by Francis Bacon (1561–1626) as more creative and uniquely capable of leading to new discoveries. Essentially, a way of arriving at universal propositions ('all wood floats') from collective sensory experiences (oak floats, sycamore floats, etc.) [50, 67, 111, 127, 180].

inertia – the tendency of a body to maintain its state of rest or uniform motion in a straight line. A concept established by Newton but prefigured in the work of Descartes, Gassendi and others. An important alternative to the Aristotelian dictum that 'everything which moves is moved by another', which implies that motion must be maintained by continuous application of force [55, 78, 233, 235].

instrumentalism – the position that scientific theories do not represent reality but are merely instruments which enable us to make predictions about natural events and processes. Most famously represented in our period by Andreas Osiander (1498–1522), who saw Copernicus's *De Revolutionibus Orbium Coelestium* (1543) through the press only after adding his own unauthorized instrumentalist preface [126, 127, 238].

intellectualist theology – emphasizes the role of God's reason in the act of Creation and implies that God was led by his reason to create the best possible world, in which nothing is contingent but based upon eternal principles of goodness, truth and reason. It follows that it should be possible to discover the system of the world by the use of reason. Compare with voluntarist theology [15, 140, 160, 164].

internal history – history writing which concentrates on technical developments in science with little or no attempt to set these in the broader cultural context. Compare with external history [199].

Interregnum – denotes the period in England and Scotland between the execution of Charles I (1600–49) and the Restoration of Charles II (1630–85) in 1660.

inverse square law – shorthand way of referring to the law of gravity, which describes how attraction varies in the opposite direction to variation in the square of the distance between two bodies. So, as the square of the distance gets bigger, the attraction diminishes, and vice versa. If the distance between two bodies increases by 4 units of distance, the attractive force between them will diminish by $4^2 = 16$ units of attractive force [99, 233, 235, 236].

kinematics – the science of motion. Used here (I hope consistently) to distinguish explanations based upon the motions of bodies from explanations based upon force (dynamics). Galileo and Descartes, in particular, tended to avoid basing their assumptions on notions of force (often regarded as occult), referring merely to the motions of bodies [80, 81].

Latitudinarianism – used here to refer to the position of a major faction of the Church of England in the late seventeenth century, in which religious conflict was avoided by insisting only upon a small number of undeniable doctrines, and professing all other matters of faith to be indifferent to one's salvation. Seen as an important element in the formation of the successful methodology of English science after the Restoration [202, 243, 124, 112].

Laudian Anglicanism – form of Anglican liturgy favoured and imposed by Archbishop William Laud (1573–1645); considered by many in England to be too close to Roman Catholicism [218].

macrocosm – the world system. See also 'microcosm'.

magic, mathematical – refers not only to magic based upon the manipulation and supposed significance of numbers (as in numerology), but also to effects brought about by machinery or other hidden contrivances [61, 62, 29, 104].

magic, natural – magic based upon the exploitation of the natural powers or virtues of things to interact with other things to bring about particular effects [25, 29, 38, 39, 111, 222, 230].

magic, spiritual and demonic – magic based upon the summoning of angels or demons to do one's bidding [25, 26, 29, 230].

manifest qualities – qualities held to arise from the four elements or their combinations: primarily heat, coldness, dryness and wetness, or secondarily softness, hardness, sweetness, sourness, and other qualities which can be directly discerned by the senses [38, 121, 153].

mechanical philosophy – major new system of philosophy developed, in different versions, during the Scientific Revolution. In its strictest forms all the properties of bodies were held to derive from the shape, size, arrangement and motions of invisibly small particles, and all causation took place by contact action. Explanations were presented as analogous to mechanical models [55, 63, 83, 89, 207, 221, 228, 233, 244]. Less strict versions of the mechanical philosophy allowed for occult qualities in matter, provided they could be defended upon empirical grounds (see experimental philosophy).

mechanicism – used here to refer to belief in the mechanical philosophy.

mechanics – traditionally the theory of machines, particularly the five 'simple machines': the lever, wedge, pulley, screw and windlass. But changing during the Scientific Revolution to include theories of impact and other problems associated with moving bodies [80, 81, 137].

metaphysics – philosophical study and theory of first principles or fundamental precepts. The Aristotelian definition of an object in terms of its matter and form is a metaphysical position, as is the mechanicist claim that objects are defined in terms of conglomerations of invisibly small particles in specific combinations and arrangements [107, 84, 89].

methodology – the study and theory of the correct methods and procedures to arrive at secure knowledge of nature [50, 55, 99, 235].

microcosm – the human being (usually *man*, of course), seen as encapsulating in miniature all the complexity and diversity of the universe itself, the macrocosm. An important notion in the magical tradition; underwriting, for example, claims about correspondences between the stars and parts of the human body [53, 217, 229, 230].

minima naturalia – Aristotelian concept of a minimum size below which a substance cannot maintain its distinct form (but reverts to being undifferentiated matter). Used in proto-chemical theorizing and influential in the revival of atomist ways of thinking [221, 63].

mirabilia – literally, 'marvellous things'. Used to denote machines or automata which frequently featured in court spectacles, ceremonies, masques and similar occasions and which were intended to produce impressive or surprising but merely entertaining effects by hidden means [13, 61, 62].

mixed mathematical sciences – astronomy, optics, music, statics and other attempts to explain physical phenomena in terms of abstract mathematics were designated 'mixed' sciences by Aristotelians because they attempted to mix explanations characteristic of one science (say, geometry) with another (natural philosophy). As such, they were often held to be less certain than the pure science of natural philosophy [50, 107, 138].

moral certainty – that certainty which, given the evidence, would convince any reasonable witness. A probabilistic form of certainty invoked by new philosophers after the old Aristotelian criteria of certainty were no longer viable [201, 203].

natural philosophy – the attempt to understand and explain the workings of the natural world. Should not be seen as merely signifying what we call science, since a number of aspects of our notion of science were *not* part of natural philosophy until the Scientific Revolution. In particular, empirical and mathematical studies had to be shown to be relevant to and combined with traditional natural philosophy in the early modern period [47, 50, 81, 80].

natural theology – the study of the natural world as a means of establishing the existence, and some of the attributes, of God. Consequently, the religious position that sound theology is based upon naturalistic principles and evidence [15, 90, 115, 125, 161, 175, 234].

Neoplatonism – strictly, refers to the diverse philosophy of Hellenistic and later thinkers who saw themselves as followers of Plato (427–347 BC), but who tended to emphasize the more religious elements in his thought. An ill-defined system of beliefs but often associated with the magical world-view [56, 71, 144, 230].

numerology – the study and theory of the significance of numbers. A magical belief that numbers can be used to reveal God's purposes and plans [53, 71, 99].

occasionalism – philosophical position, deriving from Cartesian mechanicism, which holds God to be the only true cause of physical change. Motion seems to be transferred from one body to another in impact because God maintains the system in accordance with self-imposed rules or laws of motion [89, 106, 129, 143].

occult qualities – hidden properties of substances which cannot be discerned directly by the senses but only indirectly through their effects, and cannot be reduced to the operations of the manifest qualities [121, 153, 38, 39, 110].

optics – the science of light and vision. One of the 'mixed mathematical sciences' [177, 183].

ovism – the belief that the female egg is the crucial element in reproduction, being the *sine qua non* for generation of new progeny. William Harvey (1578–1657) upheld the belief that all creatures emerge from an egg but he believed that emergence took place by gradual differentiation of previously undifferentiated organic substance (epigenesis), more commonly it was supposed that creatures were preformed in the egg; an idea which gave rise to the theory of *emboîtement* [101, 235, 172].

physiology – used here in the modern sense to denote the study of the workings of the various organs of the animal body, their form, function and role in maintaining the life of the animal [101].

Platonic – derived from or based upon the work of Plato (427–347 BC).

pre-existence – belief that all potential progeny already exist preformed in the egg (ovist) or seed (animalculist) of all potential parents. Dominant belief among mechanical philosophers who found epigenetic theories hard to accommodate to their views [101, 172, 182, 235].

preformationism – theory of animal generation which assumes pre-existence of progeny in either the egg (ovism) or the male seed (animalculism) of the parents.

Ptolemaic – derived from or based upon the work of Claudius Ptolemy (*c.*100–170). Usually used to designate the astronomical system in use before Copernicanism began to take hold [43, 60, 135, 168].

Puritanism – controversial term [see 115, 158, 227, 231, 33], but usually taken to refer to more ascetic forms of Protestant belief, particularly Calvinism.

realism – belief that the sciences reveal to us the way things really are. So, for the realist, astronomy shows us that the sun really moves around the earth, or *vice versa*, and the mechanical philosophy shows us that bodies really are made up of invisibly small particles [126, 135, 238, 83, 89, 201]. Compare with instrumentalism.

Restoration – the act of restoring the monarchy after the Interregnum period of Republicanism and the Protectorate of Oliver Cromwell (1599–1658). Also used to denote the period following the Restoration.

retrograde motion – name given to the motion of a planet when it temporarily moves in the opposite direction across the sky to its normal direction. Since the acceptance of heliocentrism, known to be an illusion caused by the earth overtaking a planet, but explained in Ptolemaic astronomy by assuming that the planet loops-the-loop on its epicycle [43, 60, 135, 168].

scholastic – derived from or based upon the work of Aristotelian natural philosophers working within the university system [192].

signatures – signs or indications, imposed upon things by God, which reveal correspondences or hidden relationships with other things. The walnut

127

looks similar to the human brain, this is a signature indicating its efficacy in treating ailments of the brain [6, 38, 39, 111, 216].

sphere, heavenly – the sphere of Mars, say, is not the same as what we think of as the body of the planet. It refers to a vast sphere, completely surrounding and centred upon, or close to, the earth. The sphere of a planet has to be thick enough to accommodate the epicycle calculated for the planet by the astronomers. The epicycle and deferent are seen as geometrical constructions which analyse the motion of the planet within its sphere. The spheres were often regarded as rigid crystalline bodies, invisible except for the luminous marker provided by the planet itself. This concept could not survive the advent of Copernicanism [138, 43, 60, 135, 168].

statics – the study and mathematical analysis of weights, balances, pulleys, levers and other systems in equilibrium and so not moving. One of the mixed mathematical sciences [55, 58, 59].

sublunary – below the sphere of the moon. See also 'superlunary'.

substantial form – in the Aristotelian tradition a particular object was made up of matter and form; form giving shape to matter. The substantial form of a thing gives it its essential, defining, properties and so makes it what it is. This idea was even extended to causative explanations, bodies being said to behave the way they do because of their substantial form [63, 84, 164].

superlunary – above the sphere of the moon. An important distinction to make in Aristotelian philosophy in which, for example, natural motions were different above and below the moon [43, 135, 138, 139].

syllogism – a formalized argument of three terms: two premises and a conclusion. For example: All men are mortal, Socrates is a man, therefore Socrates is mortal. Held by Aristotle to be foundational to all deductive reasoning, but denounced by Francis Bacon as incapable of producing new knowledge (since the conclusion is always implicit in the premises) [50, 67, 127, 138, 145].

therapeutics – system of theory in medicine concerned with determining therapies or treatments for ill-health. Traditional therapeutics, deriving from ancient Greek medicine, was concerned to restore the balance of the four humours in the body [35, 101, 139, 165, 229].

vis viva – 'living force', name given by G. W. Leibniz (1646–1716) to refer to the kind of active force produced by a body actually in motion, such as the impact caused by a falling body [3, 84, 123].

vita activa – the active life, contrasted with vita contemplativa, and signifying devotion to public service, efforts liable to benefit the 'commonwealth', and so on [145].

vitalism – philosophical position opposed to mechanicism which holds that living creatures have a vital or life-giving principle [19, 101, 165].

voluntarist theology – emphasizes God's omnipotence and unconstrained freedom of will in the act of Creation. Implies that the world is entirely contingent, depending upon nothing but God's arbitrary will, and that the system of the world cannot be discovered by reasoning but only by empirical investigation [15, 110, 112, 133, 160, 164, 234].

whiggism – a historiographical position, generally lamentable, which judges the significance of past events in the light of present-day standards, pre-occupations, etc. Or which concerns itself only with those past developments which seem obviously to have led to the current state of affairs. An ever-present threat likely to compromise work in the history of science [199].

world/world system – does *not* refer to the earth but to the cosmos or universe as a whole [217].

Index

his difficulties with the Church, 59, 79–80; his religious beliefs, 60, 75–6, 78, 79–80; his theories of life, 67–8, 69–70; *Discourse on Method*, 19–20, 68; *Le Monde*, 59; *Principia Philosophiae*, 59, 61; *Traité de l'homme*, 68

Devil, the, 43, 84, 85

Diderot, Denis (1713–84), 94, 95

Digby, Sir Kenelm (1603–65), 78

Dioscorides (*fl.* AD 1st century), 29

disciplines, 4, 11, 12, 24, 30, 56, 57; discipline boundaries, 5, 35

dissection, 26, 31, 70

Dobbs, Betty Jo T., 51

Dominicans, 73

dualism, 78–9

Duke of Savoy, 15

dynamics, 67; *see also* force

Eamon, William, 46

earth, 9, 10, 11, 12, 14, 16, 17, 18, 23, 49, 59, 63, 94

earthshine, 18

Easter, difficulties of dating, 9

eclecticism, 6, 7, 58

effluvia, 53

electrical machines, 24

emboîtement, 70

Encyclopédie, L', 94

Enlightenment, the, 94–6

Ent, George (1604–89), 28

Epicurus (*c.*341–270 BC), 58, 75, 84

epigenesis, 70

Eucharist, 79, 80

evangelism, 85

experimental method, 2, 8, 10, 18, 23–41, 24, 25, 26, 27–8, 30, 32, 33, 34, 35, 36–41, 42, 43, 46, 48, 49, 51, 53, 54, 55, 61, 62, 66, 67, 68, 71, 77, 78, 81, 82, 86, 87, 88, 89, 94; in England, 38–9, 40, 54–5; on the Continent, 37–8, 39–40

exploration, 14, 26

externalism, 6

Fabricius, Hieronymus (1533–1619), 27

Farnese, Ottavio (1598–1643), 15

feedback devices, 90–2

Ferdinand II (1578–1637), 48

Fermat, Pierre de, 23

fermentation, 68, 69

Ficino, Marsilio (1433–99), 42

Fludd, Robert (1574–1637), 45, 46

Fontenelle, Bernard Le Bovier de (1657–1757), 94, 95

force, 14, 17, 20, 21, 22, 49, 51, 52, 62, 63, 64, 65, 66, 67, 75, 76, 78, 95; centrifugal, 20, 22, 63; centripetal, 22

fortification, 4, 15

Fracastoro, Girolamo (*c.*1478–1553), 53, 58

free fall, 16

Frisius, Gemma (1508–55), 23

Fuchs, Leonard (1501–66), 29

Galen (*c.* AD 130–201), 26, 27, 28, 47, 48

Galenism, 28, 47–8, 56, 67, 72

Galilei, Galileo (1564–1642), 2, 3, 4, 5, 6, 14, 15, 16–18, 20, 21, 23, 25, 35, 37; difficulties with the Church, 59, 73–4, 80, 83; his mathematical realism, 16–18; his theory of motion, 16–17; *Dialogue on the Two Chief World Systems*, 17, 73; *Starry Messenger*, 5, 17

Gassendi, Pierre (1592–1655), 52, 58, 62, 74, 75, 77, 78, 84

Geber (*c.*721–815), 52

gender, 92–3

generation, animal, 69–71

geography, 26

geology, 26

Gesner, Conrad (1516–65), 29
ghosts, 84–5
Gilbert, William (1540–1603), 13, 14, 48, 49; *De Magnete*, 49
Glorious Revolution, the, 82, 92
God, 19, 31, 43, 44, 45, 49, 50, 60, 66, 67, 74, 75, 76, 77, 78, 81, 83, 84, 85, 86, 90, 91
Goorle, David van (*fl.*1610), 58
Grassi, Orazio (*c.* 1590–1654), 37
gravity, 22, 50, 52, 54, 59, 60, 62, 63, 64, 66, 67
Great Chain of Being, 30

harmony, 23; celestial, 45, 50, 51
Harvey, William (1578–1657), 27–8, 37, 67–8, 70, 89; *De Circulatione Sanguinis*, 89; *De Generatione Animalium*, 89; *De Motu Cordis et Sanguinis* (*Disputation on the Motion of the Heart and Blood*), 27, 89
Hatfield, Gary, 18
heart, 27, 28, 67, 68, 69, 89; systole as active stroke, 67, 68; pulsific faculty of, 68
heat, 53, 54, 59, 68, 95
hedonism, 82
Helmont, Joan Baptista van (1579–1644), 33, 52, 74
Helmontianism, 33, 47
herbals, 29
Hermes Trismegistus, 42
Hermetic tradition, 42
Hero of Alexandria (*fl.* AD 1st century), 15, 90
Highmore, Nathaniel (1613–85), 28
historiography, 2, 8, 42, 81–2, 92
Hobbes, Thomas (1588–1679), 24, 61, 62, 67, 78, 89, 95
Hooke, Robert (1635–1703), 22, 23, 28, 31, 32, 51, 54
humanists, 11, 15, 26, 29, 30, 35, 84

Huygens, Christiaan (1629–95), 20, 23, 63, 65; *Discourse on the Cause of Fall* (*Discours de la cause de la pesanteur*), 63
hydrostatics, 15, 21, 68

impetus theory, 3, 18
indivisibles, 23, 40, 58, 76; *see also* atomism
inertia, 3, 20
insects, 31, 32, 70
instrumentalism, 8, 10
instruments, mathematical, 24, 25; scientific or natural philosophical, 24–5, 35, 46
internalism, 6, 7
Interregnum, 84, 88

Jardin des Plantes, 32
Jesuits, 18, 19, 37, 73, 80
Jupiter, 18, 45; moons of, 18

Kepler, Johannes (1571–1630), 2, 13–14, 17, 21, 22, 23, 44–6, 48, 49–50, 51, 74; his mathematical realism, 13–14; his interest in occult traditions, 44–6, 48, 49–50; his religous beliefs, 44, 45, 49, 50, 74; *Astronomia Nova*, 13, 49
kinematics, 2, 4, 15, 16
kingship, forms of, 67, 86
knowledge, scientific, 1, 4, 8, 19, 23, 24, 25, 26, 27, 30, 31, 36, 37–8, 40, 43, 46, 47, 48, 61, 73, 81, 88, 92, 96
Kuhn, Thomas, 10

land reclamation, 15
latitude, 25
Latitudinarianism, 82
laws, of motion (Cartesian), 21, 60, 64, 75; of motion (Newtonian), 21; of planetary motion (Keplerian), 14, 21, 22, 50, 51; of nature, 2, 20, 39, 69, 70, 76, 87–8, 96; inverse square law, 22, 51